André Michaux in Florida

University Press of Florida

Florida A&M University, Tallahassee
Florida Atlantic University, Boca Raton
Florida Gulf Coast University, Ft. Myers
Florida International University, Miami
Florida State University, Tallahassee
New College of Florida, Sarasota
University of Central Florida, Orlando
University of Florida, Gainesville
University of North Florida, Jacksonville
University of South Florida, Tampa
University of West Florida, Pensacola

André Michaux in Florida

An Eighteenth-Century Botanical Journey

Walter Kingsley Taylor and Eliane M. Norman

University Press of Florida
Gainesville · Tallahassee · Tampa · Boca Raton
Pensacola · Orlando · Miami · Jacksonville · Ft. Myers · Sarasota

Copyright 2002 by Walter Kingsley Taylor and Eliane M. Norman
All rights reserved
Published in the United States of America

First cloth printing, 2002
First paperback printing, 2023

28 27 26 25 24 23 6 5 4 3 2 1

Library of Congress Cataloging-in-Publication Data
Taylor, Walter Kingsley, 1939-
André Michaux in Florida: an eighteenth-century botanical journey/Walter Kingsley Taylor and Eliane M. Norman.
p. cm.
Includes bibliographical references (p.).
ISBN 978-0-8130-2444-8 (cloth) | ISBN 978-0-8130-8045-1 (pbk.)
1. Michaux, André, 1746-1802. 2. Botanists—France—Biography.
3. Plant collectors—France—Biography. 4. Plant collecting—Florida—History. 5. Botany—Florida—History. I. Norman, Eliane M. II. Title.
QK31.M45 T38 2002
580'.92—dc21
[B] 2001054024

The University Press of Florida is the scholarly publishing agency for the State University System of Florida, comprising Florida A&M University, Florida Atlantic University, Florida Gulf Coast University, Florida International University, Florida State University, New College of Florida, University of Central Florida, University of Florida, University of North Florida, University of South Florida, and University of West Florida.

University Press of Florida
2046 NE Waldo Road
Suite 2100
Gainesville, FL 32609
http://upress.ufl.edu

Contents

Figures

Photos

Maps

Preface

> "Good God! when I consider the melancholy fate of so many of (botany's) votaries I am tempted to ask whether men are in their right minds who so desperately risk life and everything else through their love of collecting plants."
>
> Linnaeus 1737 (Tyler-Whittle 1997)

Shortly after his arrival in New York on November 13, 1785, and until his departure for France on August 13, 1796, André Michaux (1746–1803) explored nearly every eastern state of the new nation from New York to Georgia, to the Mississippi River, to the Hudson Bay area of Canada, the Bahamas, and Spanish East Florida. Michaux was on assignment from the French monarchy (the French Republic in 1792) to study, collect, and ship seeds, fruits, and living plants from North America to France, where they would be used to reforest the French countryside and royal nurseries and grounds of the royal park, Rambouillet. In addition to plants, Michaux shipped preserved skins and living specimens of birds and other animals to his beloved France.

Michaux's assignment to the New World was due to his fortitude, enthusiasm, and productivity displayed during his previous travels: England (1779), the mountainous French province of Auvergne (1780), the Pyrénées and northern Spain (1780), and the deserts of southern Turkey, Iraq (Mesopotamia), and Iran (Persia) (1782–1785). In addition to these successful explorations, the recommendation Michaux received from his mentor, Louis-Guillaume Le Monnier, King Louis XVI's personal physician, certainly had an influence in his getting the New World assignment (Lacroix 1938). In July of 1785, just before departing France for New York, Michaux had been honored with the title of botanist to the king. This prestigious position gave him a status similar to that of ambassadors, ministers, consuls, and other French diplomats.

For eleven years André Michaux roamed "les wilderness" of North America. His fifteen-year-old son, François André, accompanied him for the first four and one-half years of explorations. In 1790, François returned to France to complete his education. When John and William Bartram completed their explorations, seeking plants and other natural historical treasures of America, neither had traveled as extensively as André Michaux did during his eleven-year sojourn. Yet, the ambitious Frenchman is not as well known today to scientists and laypersons. He has not received the acclaim for his accomplishments in North America that he so richly deserves.

André Michaux kept a *Journal of My Voyage* (hereafter called *Journal*), where he recorded his daily activities, that provides valuable insights into his life and travels. The *Journal* was contained in ten notebooks, or cahiers, each about 10 x 15 cm in size (photos 1 and 2). In 1824 François André donated his father's notebooks, except the first notebook that contained

Photo 1. André Michaux's 3rd notebook (cahier) of his *Journal*. The cahier, about 10 x 15 cm, contains the Florida account.

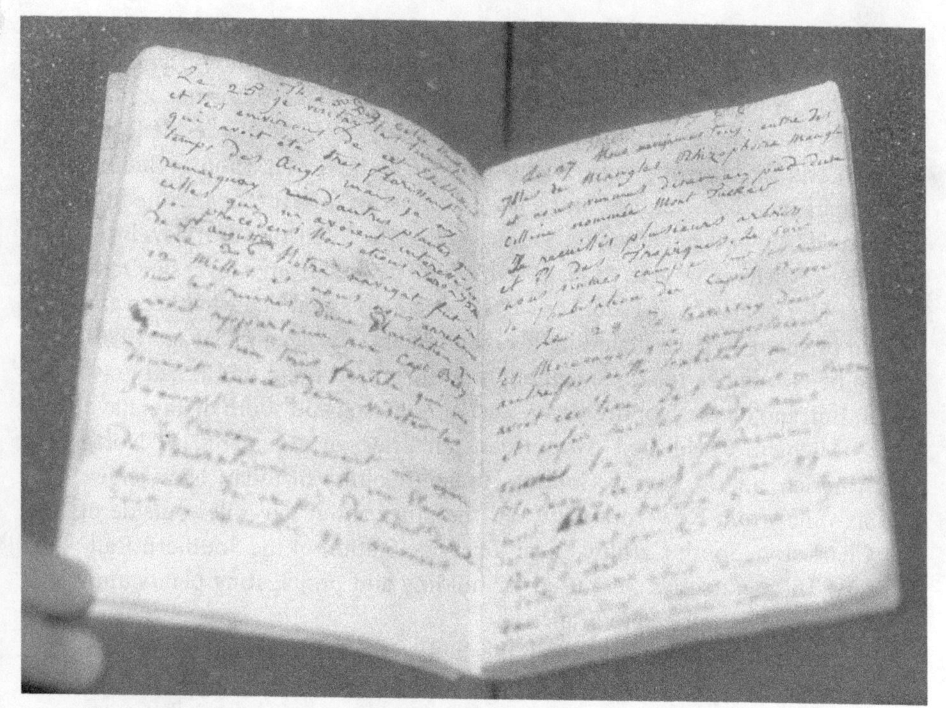

Photo 2. Open view of André Michaux's 3rd cahier. On the right page is Michaux's mention of Mt. Tucker (Turtle Mound of today), Volusia County, Florida.

the beginning accounts of his American explorations, to the American Philosophical Society in Philadelphia, where they remain to this day. The notebook containing the first part of his *Journal* was lost on October 9–10, 1796, in a shipwreck that occurred off the coast of Holland when Michaux was returning to France from America.

André Michaux's plant collection survived the shipwreck and is preserved and housed in the herbarium of the Muséum National d'Histoire Naturelle in Paris. The collection has been photographed and made available as a microfiche collection, no. 6211, by the Inter Documentation Company (IDC 1968) in the Netherlands. The IDC collection has been used extensively in this present work. The system of Uttal (1984) has been followed; each specimen on a microfiche film card has an assigned IDC number.

One important product that resulted from Michaux's studies and travels in North America was the publication in 1803 of his *Flora Boreali-Americana*—the first systematic account of North American plants. Michaux had left Paris in October 1800 as a member of Baudin's Expedition to explore New Holland, today Australia. He placed the final stages of preparation for publication of the *Flora* in the hands of his son François André, who was assisted by Louis Claude-Marie Richard, former student of Bernard de Jussieu and professor of botany at the Ecole de Médecine in Paris. Because of disagreements with Captain Nicolas Thomas Baudin, Michaux, most scientists, and a large number of the crew abandoned the expedition and remained on the Île de France, Mauritius of today. From there, Michaux went to Madagascar, where he died from a fever on October 11, 1803 (Lacroix 1938, Dorr 1997). Michaux's death date has been given by many writers, including his biographers (Savage and Savage 1986), to be November 1802. He is thought to be buried in an unmarked grave at Isatrano on the Ivondro River, west of Mahasoa and south of Tamatave.

Michaux established two garden-nurseries, or *pépinières*, in North America: one in Bergen County, Hudson County of today, New Jersey, six miles from New York City, and the other about ten miles outside of Charleston, South Carolina, near Ten Mile Station of the Southern Railway. These nurseries were used for holding and propagating plants until they could be shipped to France.

The only book that has been published on the Michaux, father and son, is that of Henry and Elizabeth Savage, *André and François André Michaux* (1986). Prior to this book, two unpublished, extensive works had been written: Léon Rey's (1954) "Deux botanistes français aux Etats-Unis: Les missions de Michaux Père et Fils (1785–1808)" and a Master of Arts thesis, "André Michaux et son exploration en Amérique du Nord, de 1785 à 1796, d'après les sources manuscrites," by Marie-Florence Lamaute (1981), then at the University of Montreal. The obituary, "Notice historique sur André Michaux," by Joseph Philippe François Deleuze (1804) is one of the earliest biographical accounts written about Michaux. Deleuze (1753–1835) was a naturalist and later librarian at the Muséum National d'Histoire Naturelle in Paris. Michaux was both friend and colleague to Deleuze. Stafleu and Cowan (1981), in *Taxonomic Literature*, also provide a wealth of bibliographic information on Michaux. Scientific papers, based on Michaux's North American explorations, that reflect a geographic perspective have been written for the Carolinas (Rembert Jr. 1979; Seaborn 1976), the Midwest (Thwaites 1904), and Canada (Brunet 1861, 1864; Dutilly and Lepage 1945; Rousseau 1948).

André Michaux's travels and botanical investigations in Florida are little known. For these reasons, we have written this in-depth account of Michaux's Florida explorations. All of the plants known to us that Michaux saw or collected while in Florida, as well as an account of the historical events, places, and people that are mentioned in his *Journal*, are presented and discussed. Furthermore, we hope that both scientists and laypersons, through this book, will become better acquainted with this remarkable Frenchman who explored North America at an important time of American history. If we could see today the forests, rivers, streams, hills, mountains, and valleys that Michaux's eyes captured, we might not recognize our own America.

After the three-month Florida trip was completed, Michaux and son returned to the Carolina pépinière. Subsequent explorations in America, before Michaux returned to France in 1796, took him to the Georgia mountains, the Bahamas, the Carolina mountains, the northeastern states of Maryland, Virginia, and Pennsylvania, to Canada's Hudson Bay area, to the states of Kentucky, Tennessee, and Illinois, and to the Mississippi River.

Acknowledgments

Writing a book that crosses many disciplines has required the cooperation and assistance of historians, botanists, biologists, linguists, archivists, librarians, and many others. We could not have finished this work had we not had the assistance of those listed below. To each, we are extremely grateful for their input in this study, no matter how great or small.

Librarians: Marion Crist, Mary Cross, Scott DeHaven, Rita Dockery, Roy Goodman, Elizabeth Carroll-Horrocks, Alison Lewis, and Eleanor Roach, American Philosophical Society Library; M. Ducreaux and Pascale Heurtel, Bibliothèque Centrale du Muséum National d'Histoire Naturelle, Paris; Christine Campbell, The British Library; Jeffrey M. Flannery, Library of Congress; Bruce Chappell, P. K. Yonge Library at the University of Florida; Charlie Williams, Public Library of Charlotte and Mecklenburg County, North Carolina; Anne Marie Allison, Joseph Andrews, Deidre Campbell, Peter Spyers-Duran, Virginia Farmer, Jeff Franks, William Fidler, Ingrid C. Hunt, Chang C. Lee, the late June Stillman, Linda Sutton, Winnie Tyler, John Walters, and Jack Webb, University of Central Florida Library; Gertrude F. LaFramboise and Kathleen J. Reich, Archives and Special Collections, Rollins College; Daniel Boice and Henry G. Fulmer, South Caroliniana Library; Sheherzad Navidi and Charles P. Tingley, St. Augustine Historical Society; Peter Shipman, Stetson University; and Ellen Cohn and Karen Duval, The Franklin Papers, Yale University Library. Botanists and other individuals associated with academic institutions are as follows: William Dress, Peter Fraissinet, and Sherry Vance, Cornell University; William L. Culberson, Duke University; Steve Glassman, Embry-Riddle University; Elizabeth Baker, Halifax Historical Society; James Hardin, North Carolina State University at Raleigh; J. Perry Edwards and Ronald L. Stuckey, The Ohio State University; Terry Farrell, Stetson University; Karl-Heinrich Barsch, José B. Fernandez, and Jerrell H. Shofner, University of Central Florida; Barbara Purdy, University of Florida; Marie-Florence Lamaute and Roland Lamontagne, University of Montreal; Bruce Hansen and Richard P. Wunderlin, University of South Florida; Neil A. Harriman, University of Wisconsin at Oshkosh; Robert Kral, Vanderbilt University; and Leonard J. Uttal, Virginia Polytechnic Institute and State University. Individuals from museums, botanical gardens, and state and federal agencies who have assisted us are Claudine Pouret, Académie des Sciences, Paris; Jan Barber, Alfred E. Schuyler, and Carol M. Spawn, Academy of Natural Sciences at Philadelphia; Florence Clavaud, Archives de France; Monique Constant, Archives du Ministère des Affaires Etrangères, Paris; Pedro González Garcia, Archivo General de Indias, Seville, Spain; Richard Howard, Arnold Arboretum; Joel T. Fry, Bartram Garden; Paul Hiepko, Botanischer Garten und Botanisches Museum Berlin-Dahlem; Ivor Kerslake, British Museum; John Stiner, Canaveral National Seashore; Marion Smith, Division of Historical Resources, Tallahassee; Joan Morris, Florida State Archives, Tallahassee; Alice James, Ingrid P. Shields, and Edward Weldon, Georgia Department of Archives and History, Atlanta; Bertil Govers and Xylander Kroon, Inter Documentation Company, Leiden; Harry E. Luther, Marie Selby Botanical Gardens; the late Joseph Ewan and Marshall Crosby, Missouri Botanical Garden; Serge Barrier, Jean-Claude Jolinon, and Marc Pignal, Muséum National d'Histoire Naturelle, Paris; Tim Padfield, Public Record Office, London; Juan Armada, Real Jardín Botanicó, Madrid; and Laurence Dorr and Joan Nowicke, Smithsonian Institution.

Those who assisted with translations are Arlie Amadeo (French), Julia M. Cabrera (Spanish), Philip Crant (French), Patricia Cruz (Spanish), Carlos Diez (Spanish), Arlette Lubin (French), Eliane M. Norman (French), Ralph Robinson (French), and Winnie Tyler (French). Others who deserve our thanks are Sherry Bufano, Francis P. Chinard, Loretta Graen Coyne, Phil Edwards, Barbara Erwin, Justin Grevich, Mike Lueg, Jerry Millen, Hugh Morton of Grandfather Mountain, North Carolina, William Reed of William Reed Company, Phillip Russell, the late Elizabeth Savage, Margaret Shannon, Sid Seymour Taylor, David H. Vickers, Charles

A. Walker Jr. of the Heritage Rose Foundation, Raleigh, North Carolina, and Vera Zimmerman.

We thank Dan Schafer, University of North Florida, Ronald L. Stuckey, The Ohio State University, and David Taylor of Converse College, whose reviews of the manuscript were appreciated. Ronald L. Stuckey, Terry Thaxton, University of Central Florida, and Daniel B. Ward, University of Florida, read earlier drafts of the manuscript and we are grateful to them. Walter Judd, University of Florida, assisted with the verification of Michaux's plants in the Ericaceae. Richard H. Jackson skillfully drew the map of St. Augustine (map 14). We thank Matt Crispell very much for drawing the city gate (figure 2) and Jodi Doster for the scene with the dugout canoe (figure 3). Eliane Norman took the photograph of *Halesia carolina* (photo 15), Jerry Millen the photograph of the mouth of Pellicer Creek (photo 16), Todd Campbell the photograph of Turtle Mound (photo 23), and Boyd Z. Thompson the photograph of Lake George (photo 30). The remaining photographs were taken by Walter K. Taylor. Bertil Govers of Inter Documentation Company granted permission to use Michaux's plants. Karin S. Taylor graciously supported and assisted us in many ways. Special thanks are extended to Kenneth Scott, director of University Press of Florida, and all those who played a role in making this book possible. Specifically we acknowledge Bennie Watson, Meredith Morris-Babb, Deidre Bryan, Lynn Werts, David Graham, Jim Denton, and our editor, Gillian Hillis, a very gracious lady.

Last, but certainly not least, we acknowledge in memory André Michaux and his son, François André, for their great contributions to our knowledge of the North American flora. This study was funded in part by a grant from the Michaux fund of the American Philosophical Society, Philadelphia. To that organization we extend our gratitude for financial support, assistance in obtaining documents, and encouragement in the project.

1

Introducing André Michaux and His Times

In the middle of the summer of 1785, Benjamin Franklin (1706–1790) extended his last farewells to his French friends and departed Paris for home. This long-time, esteemed resident of Philadelphia had completed his mission. He had been in Paris since the spring of 1782, serving as an American delegate to the peace negotiations between Great Britain and the United States. Later that year, John Adams (1735–1826) and John Jay (1745–1829) had joined Franklin in the negotiations (Hicks 1957). The surrender of British General Charles Cornwallis (1738–1805) on October 19, 1781, at Yorktown had terminated the Revolutionary War, but the final treaty (Treaty of Versailles) between Great Britain and the United States was not signed until September 3, 1783, in Paris.

For the French people, Franklin's departure was not a happy event; they hated to see their beloved patron saint leave them. Thomas Jefferson (1743–1826), Franklin's successor as minister to France, had already arrived in Paris in August 1784, having embarked on the *Ceres* from Boston on July 5 (Lisitzky 1933). Earlier, Jefferson had been scheduled by Congress to go to France as minister to assist Jay and Franklin in concluding the Treaty of Versailles, but because of difficulties encountered, the mission was aborted.

Jefferson brought with him his newly written, Congress-approved, "Instructions" for American ambassadors located abroad at all foreign capitals. He joined John Adams and Franklin as commissioner to negotiate a trade treaty with Europe. Franklin and Jefferson had nearly a year together in Paris before the former returned to Philadelphia.

During his five-year stay in France, Jefferson had contacted André Thouin (1747–1824) with the goal of exchanging seeds and plants between America and France (Ewan 1969). Thouin was head gardener of the Jardin du Roi (Jardin des Plantes after the birth of the Republic in 1792) in Paris, professor of horticulture at the museum in 1793, student of Bernard de Jussieu (1699–1777), and foremost authority on naturalization of foreign plants. A long-lasting relationship developed between Thouin and Jefferson (Savage and Savage 1986). Jefferson also met the great French naturalist Georges-Louis Leclerc, Comte de Buffon (1707–1788). The nearly eighty-year-old author of the famed forty-four-volume *Histoire Naturelle* was still writing his classic work. Jefferson witnessed the beginning of the French Revolution, and on July 17, 1789, he inspected the Bastille, which recently had been taken by the people of Paris. In October of that year Jefferson sailed home for Virginia, only to find that President George Washington (1732–1799) had appointed him secretary of state (Lisitzky 1933). Jefferson formally accepted the position in February 1790.

In late September of 1785 four Frenchmen left Port l'Orient, Lorient of today, in Brittany on the western coast of France, for New York. These men were not political delegates like Franklin and Jefferson, but represented the French monarchy under King Louis XVI (1754–1793). The prime figure of the group was André Michaux (1746–1803), a man with much less visibility, familiarity, and esteem than either Franklin or Jefferson. Both men, however, would become acquainted with Michaux in the days ahead. In an unpublished letter, dated August 9, 1785, located in the Archives du Ministère des Affairs Etrangères, Benjamin Franklin was informed by Charles Gravier, Comte de Vergennes (1719–1787), minister of foreign affairs under Louis XVI, of Michaux's visit to America (appendix 2-3). The letter of introduction, which Michaux was to present to Franklin, informed the patriot of his assignment and requested Franklin's assistance. Franklin was told that Michaux represented the King of France.

Benjamin Franklin returned to Philadelphia, the future capital of the American colonies and hub of political activities. Here the foundations of American government would be constructed and honed with the opening of the Constitutional Convention on May 25, 1787. Philadelphia also became the initial center for the development of North American botany, in part because of the accomplishments of John Bartram (1699–1777) and his son, William (1739–1823), and the 1731 founding of the renowned Bartram Garden, the first of its kind. Another contributing factor was the appointment to the University of Pennsylvania in 1789 of Dr. Benjamin

Photo 3. Home of John and William Bartram located west of Philadelphia on the Schuylkill River. The famous Bartram Garden surrounds the home.

Smith Barton (1766–1815), physician-naturalist-professor, whose extensive library, herbarium, and expertise in materia medica and natural history attracted many students interested in the native American flora and medicine (Reveal and Pringle 1993).

John Bartram had been dead for eight years, but Benjamin Franklin, aged 79, continued to visit the Bartram home and Garden located west of Philadelphia on the Schuylkill River (photo 3). Both men were among the nine original members of the American Philosophical Society, a prestigious organization that still exists, modeled on the Royal Society of London (Reveal 1992).

Most of Bartram's generation of pioneer-naturalists and close associates who lived in America were now deceased. Cadwallader Colden (1688–1776), a wealthy politician, physician, and naturalist, had come to America from England in the early 1700s. He settled in New York where he became surveyor general and governor. Colden's large estate, Coldengham, flourished with impressive gardens. His friend, John Clayton (1686–1773), also of England, had settled on a productive plantation in Virginia around 1715. Clayton was clerk of court for Gloucester County, collected plants for forty years, and compiled a catalog of plants of that area that became the basic material for *Flora Virginica* (first volume published in 1739; second volume in 1743, Leiden), authored by the Leiden physician and botanist Joannes Fredericus Gronovius (1690–1762). Gronovius used Clayton's plants without his permission (Reveal 1992). Even the second edition of the *Flora*, published in 1762, carried Gronovius as author. John Mitchell (1711–1768), native of Lancaster County, Virginia, studied medicine at the University of Edinburgh, Scotland, and returned home, where he practiced at Urbanna, Middlesex County, Virginia (Berkeley and Berkeley 1982). Like Bartram, Mitchell was a good friend of Clayton. Mitchell studied the colony's flora and fauna until 1746, when he returned to England because of poor health. Here he spent the last twenty-two years of his life (Berkeley and Berkeley 1974).

The only member of this quintet still alive in 1785 was the youngest of the associates, Dr. Alexander Garden (1730–1791). This Scottish native came to America in 1751 and emigrated in 1752 to Charles Town (Charleston), South Carolina. He, too, had studied medicine at the University of Edinburgh. Garden spent most of the prime of his life at Charles Town where he practiced medicine and botanized until forced to return to England. His loyalty had been with George William Frederick (1738–1820), better known as George III, during the Revolutionary War. Today, Garden is commemorated by the Old World genus with fragrant, white flowers, the gardenia. All of these men, though of varied backgrounds, were united by a common bond of love for nature, knowledge, and the land (Kastner 1977).

In Sweden, the eminent guru of taxonomy and prince of botany, Carolus Linnaeus (1707–1778), his champion in Holland, Joannes Fredericus Gronovius, and Linnaeus's favorite student, Pehr (Peter) Kalm (1715–1779) of Sweden, were also dead. Linnaeus survived John Bartram by five months. The Swedish naturalist never visited the United States, but Kalm did. In September 1748 and several times thereafter, Kalm visited John Bartram and Franklin (Berkeley and Berkeley 1982). Before leaving the United States in October of 1750, Kalm had botanized in much of the Northeast and as far west as the Great Lakes (Kastner 1977; Reveal and Pringle 1993).

In England, Peter Collinson (1694–1768), the well-to-do London wool merchant, avid gardener, fellow Quaker, and correspondent of John Bartram for thirty-five years, was also dead. It was Collinson who made John Bartram a household name among early naturalists, especially to the scientific world of Europe. Collinson engaged Bartram to send American seeds and plants to him and to others in the Old World. Collinson introduced the name of John Bartram to Linnaeus, who paid Bartram a high tribute for his botanical endeavors when he called him the "greatest natural botanist in the world" (Cheston 1953). Collinson also was instrumental in John Bartram's becoming King's botanist to George III in 1765. Bartram held this title, along with the annual salary of fifty pounds, until the American Revolution began (Berkeley and Berkeley 1982). Peter Collinson was responsible, more than any other person, for introducing American plants to Europe's botanists and gardeners. Neither Collinson nor Linnaeus ever met the senior Bartram.

In Paris, the renowned botanist Bernard de Jussieu was dead, but his nephew, Antoine-Laurent de Jussieu (1748–1836), brought to fruition his uncle's ideas on a natural system of plant classification in his 1789 *Genera Plantarum* (Stevens 1994). Jean-Baptiste-Pierre-Antoine de Monet de Lamarck (1744–1829) and Charles-Louis L'Héritier de Brutelle (1746–1800), member of l'Académie des sciences and director of the Plant Conservatory in Paris, were actively engaged in their studies. Lamarck, a gifted botanist and author of the three-volume *Flore Françoise* (1778) and thirteen-volume *Encyclopédie Méthodique: Botanique* (1783–1817), is remembered also for his work on invertebrate taxonomy and ideas on evolution.

Nearly nine years had passed since William Bartram had returned home on January 1, 1777, from his famous trek of nearly four years throughout the Southeast, only eight months before his father's death on September 22 (Harper 1958; Slaughter 1996b). Despite John Bartram's title of King's botanist, recognition from Linnaeus and other prominent naturalists of the time, publication of his journal account describing the East Florida explorations in 1765–1766, and numerous other accomplishments, it was William who has perpetuated the Bartram name to the present. With the publication in 1791 of his widely acclaimed book *Travels through North and South Carolina, Georgia, East and West Florida*, William firmly stamped the Bartram surname in the everlasting annals of North American natural history. He was the first native-born American to devote his entire life to the study of nature (Slaughter 1996a).

In 1785, unmarried, William Bartram, then forty-six, never undertook another major exploration, though he was offered by Jefferson the position of natural history advisor to the Thomas Freeman expedition of 1806 up the Red River (Cheston 1953). He also never assumed the professorship of botany offered by the trustees at the University of the State of Pennsylvania, now University of Pennsylvania. Instead, William spent the rest of his days at the Bartram home, writing, drawing, assisting his younger brother John Jr. (1743–1812) in the business affairs of operating and caring for the Bartram Garden, and in greeting numerous visitors from the fields of science, literature, and politics who came to see the place and its curiosities. People from all areas visited the famous traveler and the Garden including Benjamin Franklin, Thomas Jefferson, James Madison (1751–1836), Alexander Hamilton (1757–1804), George Mason (1725–1792), William Darlington (1782–1863), Dr. Benjamin Smith Barton, Dr. Benjamin Rush (1746–1813), The Rev. Manasseh Cutler (1742–1823), and George Washington. Through these exchanges, William Bartram became mentor to many naturalists seeking botanical treasures yet to be discovered in the hinterland of America.

One such visitor who made at least seven visits to the Bartram Garden was the intrepid French botanist André Michaux. William Bartram complimented Michaux, seven years his junior, when he said that Michaux was one of the very few collectors who could traverse the ground that he and his father had covered and find plants that they had missed (Kastner 1977). Other writers have bestowed similar accolades on André Michaux: "one of the most assiduous plant collectors in eastern U.S." (Morton 1967), "one of the most important personalities in the botanical explorations of the 18th century in America" (Rembert, Jr. 1979), "a man of unusual intelligence" (Thwaites 1904), "a formidable figure" (Kastner 1977), "indefatigable botanist and persevering traveller" (Gray 1882), "a botanist of merit and of a special quality" (Lacroix 1938), "indefatigable" (Nuttall 1822), "eminent—the name ever illustrious in the annals of American botany" (True 1937), and "the greatest—finest botanist to dig *Rosa laevigata* in Georgia" (Shannon 1983). Joseph Ewan (1965), the dean of botanical historians, aptly wrote that "None of the French naturalists accomplished so much as Michaux and his son François André in discovering and reporting on our native plants."

André Michaux was an achiever of firsts—the first trained botanical explorer to travel extensively in North America, first botanical explorer of the higher Allegheny Mountains (Gray 1882), first botanist to venture to the Hudson Bay area in Canada to collect specimens (Kastner 1977), farthest traveled and best trained scientist to enter the American community of naturalists at that time (Kastner 1977), first naturalist to see the North American prairie, and author of the first North American flora, *Flora Boreali-Americana* (Michaux 1803). The *Flora* contains the first inventory of North American weeds. Unlike the Bartrams, Michaux made extensive collections of grasses (Hitchcock 1908), sedges (Core 1936), ferns (Morton 1967), mosses, and lichens. All plants that appeared in the *Flora* were collected or observed by Michaux, a remarkable feat in itself. The *Flora* confers the highest credit on the industry and acuteness of Michaux (Hooker 1825). One would be hard put to find another person who exceeded André Michaux in his tremendous love for plants, the compelling drive to find them, the number of plants and seeds collected, the number of species described, the number of plants named in his honor, and the devotion and loyalty to his native land. With his keen eye, André Michaux set straight many questionable identifications (Kastner 1977). He did not hesitate to question Linnaeus's determinations. It is abundantly clear that Michaux was well acquainted with the contents of Linnaeus's *Species Plantarum*. He renamed species previously named by Thomas Walter (1740–1789) and William Bartram. Michaux was familiar with Catesby (1771), as indicated in his *Journal*, where he noted finding in Florida a mangrove with fruits like those of Catesby's fig. Also in a letter dated April 28, 1789, to Count d'Angiviller, Michaux wrote that he observed in the Bahamas nearly all of the trees cited by Catesby.

Historians writing about the natural history of the Southeast often mention John or William Bartram and their explorations; however, they have ignored André Michaux, whose botanical discoveries exceed those of either Bartram. Why this has happened is not known. Today, there is no Michaux Club, no Michaux Association, nor a Michaux Trail similar to the Bartram Trail Society and John Bartram Association. We do not know what the Frenchman looked like. Michaux's bust was to have been placed on the façade of the temperate greenhouse at the Jardin des Plantes, along with the busts of Philibert Commerson (1727–1773), French botanist who explored Uruguay, Argentina, and Madagascar, Joseph Dombey (1742–1794), French physician and explorer of Peru, Chile, and Brazil, and other explorers, but this was never achieved (Deleuze 1804).

Perhaps worse, incongruous and unfounded statements by authors of books and scientific papers reflect a total lack of knowledge and appreciation of the man and his accomplishments. Historical writer Ambrose (1996) misconstrued André Michaux's relation with Thomas Jefferson, then secretary of state, involving a 1793 plan that was devised by the Girondist Ministry in Paris and to be executed by Edmond Charles Genêt (1762–1834), the young French foreign minister to the United States. Jefferson, with support of members of the American Philosophical Society, had chosen Michaux to explore the country between the Mississippi River and the Pacific Ocean—the exploration that became the Lewis and Clark expedition in 1803. But with the appearance of Genêt on American soil, Michaux's plans were diverted and Jefferson had to wait another decade for his westward expedition to be achieved.

Genêt's intent was to liberate Louisiana from Spanish control by performing a joint attack with France's naval force and American volunteers raised in Kentucky and placed under the command of George Rogers Clark (1752–1818), brother of William Clark (1770–1838) of the Lewis and Clark expedition. Michaux accepted the commission as Genêt's liaison. Ambrose (1996) incorrectly writes that Michaux was a "secret agent of the French Republic whose chief aim was to raise a western force to attack Spanish possessions beyond the Mississippi. At Jefferson's insistence, the French government recalled Michaux." The chief aim of Michaux was to discover the botanical treasures of "les wilderness" and not to be a secret agent for France. Thomas Jefferson was aware of the projected liberation of Louisiana and initially gave his full support to the operation. Jefferson never insisted that Michaux be recalled by the French government. The Washington administration did ask for the recall of Genêt, who would have faced the guillotine upon his return to France (Catanzariti 1992). In March 1794, Genêt's commission was canceled by Jean Antoine Joseph Fauchet, his successor, who had arrived from Paris in January 1794.

George Washington changed his mind concerning Genêt's recall and extended forgiveness to him (Eifert 1965). Edmond Genêt remained in America, where he married New York Governor George Clinton's (1739–1812) daughter, Cornelia (Lisitzky 1933). Genêt died at Jamaica, Long Island, in 1834 (Thwaites 1904). With the collapse of the Genêt project, André Michaux returned to Charleston in March 1794 and continued his botanical explorations of North America until he returned to France in 1796.

André Michaux chose a simple way of life. He was willing to undergo many hardships to do his job to the very best of his ability. He was frank, impatient, and terse, and he took little part in meaningless conversation (Deleuze 1804). He was intolerant of foolishness, laziness in people, or talk that was not productive. Loss of time was totally unacceptable. He pushed his body to the limits, and had amazing perseverance and stamina. His working long hours in the tropical heat of Madagascar and disregarding the advice of fellow workers, who knew the harshness of that area's climate, probably contributed to Michaux's demise. Heat, rain, snow, frozen ground, insect pests, snakes—which he did not like—and other wild animals did not deter André Michaux from performing his work and accomplishing his mission in America. Traveling, working, and writing in his *Journal*—by light of the moon or sun, it made no difference to him.

Michaux believed in God who protected him, although he was not overly pious. He was devoted to his son, and his son held his father in high esteem. Michaux was a man of integrity and high morals who gladly assisted those in need. He refused to incur debt and repaid, whenever possible, his hosts for their hospitality and graciousness, be it for food, a place to sleep, or loan of a horse. Payment often was in the form of plants—the only thing he had to offer.

Two traits of André Michaux consistently stand out in his own writings and in accounts written by others: his passionate love of plants and his loyalty to his beloved country. He wanted his botanical work to benefit mankind (Deleuze 1804). In a letter written from Charleston on November 6, 1787, to his friend André Thouin, Michaux expressed the desire for France to be the first to publish botanical discoveries in America, before the English (appendix 2-8). He wrote that there were "two ignorant men" sent by nurseries to collect and send plants to Sir Joseph Banks (1743–1820) of England. One of the two was the Scots botanist John Fraser (1750–1811) and the other was probably Fraser's companion Dr. Porter (Coats 1969). On May 29, 1787, at Augusta, Georgia, Michaux wrote in his *Journal* of a "certain Mr. Fraser" who had accompanied him since their departure from Charleston. "Fraser," wrote Michaux, "lacked resources and knowledge in natural history and in the botanicals. He wasted precious time collecting many large plants of no value." Michaux further expounds that he had become exhausted by Fraser's questions, his ignorance, and his lack of confidence. Having lost his horses twelve miles before reaching Augusta, Michaux told Fraser to continue his explorations and not wait for him because he had to find his horses. Fraser heeded Michaux's advice and departed, much to the Frenchman's relief. Michaux did not recover his horses.

Fraser, a born collector, made several visits to North America between 1780 and 1810 looking for and collecting plants. In 1799 he was accompanied by his son, John, who made at least two more trips to America after his father's death in 1811 (Coats 1969). Fraser was instrumental in editing Thomas Walter's *Flora Caroliniana* and getting it published in London in 1788. Coats (1969), citing Fraser's *Agrostis Cornucopiae* published in 1789, says that Fraser admitted his determination to surpass the French botanist in discoveries, or at least to obtain for Great Britain equal honor with France in the field of botany.

From Michaux's letter (dated November 6, 1787) to Thouin, it is clear that the botanist wanted to write a flora of North America. This is probably why he visited Florida, the Bahamas, and Canada, thereby ensuring that the geographic coverage of the flora would be as extensive as possible (Deleuze 1804). Plants growing in or near tropical regions such as Florida and the Bahamas would not have been an important commodity to the economic welfare of a country with a temperate climate like France.

Michaux also wanted to become a member of l'Académie des sciences. Writing a flora would certainly increase his chances of being elected to the prestigious organization, but he wanted to be elected without having to wait the several years it would take to put together a flora. But l'Abbé Nolin, head gardener of the Royal Nursery at Rambouillet, had complained about Michaux's poorly packed cases of plants that had been shipped from New York and Charleston during the early years. In a letter dated November 6, 1787, to his friend André Thouin, Michaux indicated his apprehension that Nolin's reproaches might have a negative effect on his being admitted to l'Académie. Michaux's fears, however, were not realized; in March 1796 he was elected a nonresident member of l'Académie des sciences, Section of Rural Economy and Veterinary Art (Lacroix 1938). As indicated in his *Journal*, he attended several meetings of l'Académie after his return to France in 1796.

Through his prolific letter writing, André Michaux kept his cohorts in France informed of his whereabouts and accomplishments. Michaux devoted his entire life to France and to the well-being of mankind through his discoveries of plants and their uses. He considered the oak the tree of all trees, and wrote *Histoire des Chênes de l'Amérique* (Michaux 1801). However, he was especially fond of the loblolly bay, *Gordonia lasianthus* (L.) Ellis, a member of the tea family that produces large white flowers in the summer and grows in wet areas of the Southeast (photo 4).

Early Life

André Michaux's parents, André Michaux and Marie-Charlotte Barbet, lived at the royal estate of Satory near the park of Versailles (Lamaute 1981). His father was a farmer and managed five hundred acres of the Satory domain of King Louis XV (1710–1774). Here on this farm the Mi-

Photo 4. Loblolly bay, *Gordonia lasianthus* (L.) Ellis, a common member of the tea family in the southeast United States. André Michaux was especially fond of this species.

chaux's son, André, was born on March 8, 1746, according to his baptismal record in the Versailles Municipal Archives (Lamaute 1981; appendix 2-1). André's first teacher of plants—how to grow and nurture them—was his father. At an early age the young André had united theory with practice in growing plants (Deleuze 1804). The young Michaux probably thought he would follow his father's footsteps.

André's early schooling ended at age fourteen, but he was trained in the classics and later studied Greek and perfected his Latin (Deleuze 1804; Savage and Savage 1986). In 1763, when André was seventeen, his father died, and three years later his mother also died.

With the death of both parents, André and his younger brother, André François, were compelled to manage the king's large farm (Lamaute 1981). André assumed the role of superintendent of the farm, but becoming a farmer was not to be his destiny. Nearing age twenty-four, André Michaux married Anne-Cécile Claye, daughter of a wealthy farmer of Beauce, in October 1769 (Lamaute 1981). In September 1770, not even a year after their marriage and soon after the birth of their son, François André (August 16, 1770–October 23, 1855), Anne-Cécile Michaux died.

André Michaux was devastated and plunged himself into his work on the farm in an attempt to overcome his loneliness and sad plight. Fortunately for Michaux and for the science of botany, Dr. Louis-Guillaume Le Monnier (1717–1799), court physician of Louis XV and of Louis XVI, physicist, superintendent of the famed Trianon Gardens, and principal professor of botany at the Jardin du Roi in Paris, became André Michaux's mentor.

Le Monnier saw in this young man a person with much potential and one who truly loved plants. He took Michaux under his wing, encouraged him, taught him, and advised him to study the naturalization and acclimatization of exotic plants (Deleuze 1804; Lacroix 1938). Le Monnier's advice was heeded. Michaux soon sought a career as naturalist-explorer-collector, a desire he had held from a very early age (Savage and Savage 1986).

André Michaux could have had no better mentor and friend than Le Monnier. He paved the way for André to begin formal studies in botany at the Trianon under the eminent Bernard de Jussieu, central figure at the Jardin du Roi, professor of botany, and brother to the famed Antoine de Jussieu (1686–1758). Le Monnier, a former student of Bernard de Jussieu, took it upon himself to introduce Michaux to Charles-Claude Flahaut de la Billarderie, Comte d'Angiviller (1730–1810). D'Angiviller was one of the most influential persons in the court of King Louis XVI, who had succeeded his grandfather, Louis XV, in 1774, and Queen Marie Antoinette (1755–1793). D'Angiviller was director and commander-general of buildings and gardens of the king, administrator of the Jardin du Roi (1788–1792) upon the death in 1788 of the great Comte de Buffon, and governor and administrator of His Majesty's chateaux and private grounds of Rambouillet, located forty km southwest of Paris, near Versailles. Rambouillet, the famed former home of Marie de Medici (1573–1642) and Henry IV (1553–1610), was the favorite residence of Louis XVI, who acquired the 33,000–acre royal park of varied terrain and intricate waterways in 1783 (Savage and Savage 1986). D'Angiviller was the main contact person in France for Michaux during his New World explorations (figure 1).

André Michaux's knowledge of botany and experience in cultivation and naturalization of plants increased. In 1779 he moved to Paris and worked at and lived near the Jardin du Roi (Deleuze 1804). Here Michaux received instructions from André Thouin and Antoine-Laurent de Jussieu (Beale 1978).

On June 1, 1779, André Michaux received from Louis XVI and Comte de Buffon, member of l'Académie des sciences and cabinet member of His

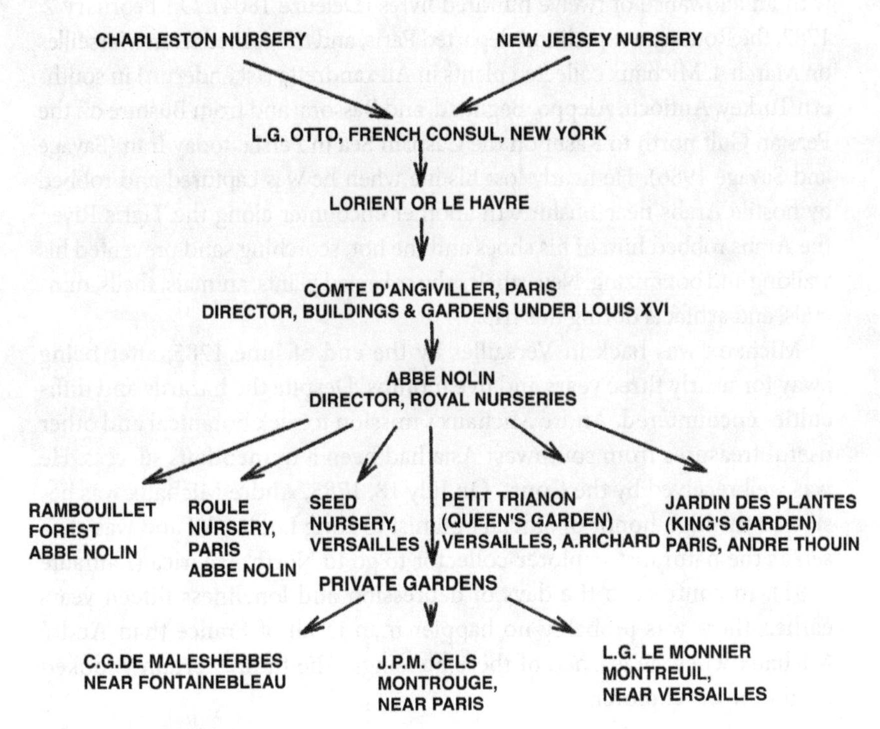

Figure 1. Protocol showing the chain of command involved in André Michaux's mission to America. Modified from Rey (1954).

Majesty, an appointment as correspondent of the Royal Garden (Lamaute 1981). Shortly thereafter, Michaux began his travels to foreign lands, a longtime dream that finally came true. First, in 1779, he visited England and the gardens there. After returning to France in 1780, Michaux, now 34, with André Thouin, age 38, and Jean-Baptiste Lamarck, age 36, visited the mountainous province of Auvergne in southern France. After that trip, the naturalists went to the Pyrenees and northern Spain for another successful trip.

After André Michaux returned from Spain he petitioned Marie Antoinette and was subsequently chosen to accompany Jean François Xavier Rousseau (1738–1808) to southwest Asia (Coats 1969). Rousseau was a French diplomat sent to Baghdad and Bassora, now Basra, Iraq; he was later consul to Persia (Savage and Savage 1986). The brother of Louis XVI, later to become Louis XVIII (1755–1824) and referred to as "Monsieur," sponsored Michaux's trip with an allowance of twelve hundred livres (Deleuze 1804). On February 2, 1782, the Rousseau expedition departed Paris, and they sailed from Marseilles on March 4. Michaux collected plants in Alexandretta (Iskenderun) in southern Turkey, Antioch, Aleppo, Baghdad, and Bassora, and from Bushire on the Persian Gulf north to Rasht on the Caspian Sea in Persia, today Iran (Savage and Savage 1986). He nearly lost his life when he was captured and robbed by hostile Arabs near Bushire. In another encounter along the Tigris River, the Arabs robbed him of his shoes and the hot, scorching sand prevented his walking and botanizing. Nevertheless he collected plants, animals, shells, minerals, and artifacts during this trip.

Michaux was back in Versailles by the end of June 1785, after being away for nearly three years and five months. Despite the hazards and difficulties encountered, André Michaux's mission to seek botanical and other useful treasures from southwest Asia had been a tremendous success. He was well received by the Court. On July 18, 1785, André Michaux was bestowed the high honor of official botanist to King Louis XVI and was chosen as the naturalist-explorer-collector to go to North America (Lamaute 1981). In contrast to the days of depression and loneliness fifteen years earlier, there was probably no happier man in all of France than André Michaux when he learned of these blessings. The future certainly looked bright for the explorer.

2

To the New World

After forty-seven days crossing stormy waters, the ship *Courrier de New York* entered New York City harbor and set anchor. She had come from Port l'Orient, on the western coast of France. It was Sunday, November 13, 1785; a frigid winter gripped New York (Michaux 1793; Duprat 1957; Savage and Savage 1986).

Among the passengers on board were four Frenchmen: André Michaux, aged thirty-nine, his fifteen-year-old son François André, Pierre Paul Saunier, aged thirty-four, gardener and student of André Thouin, and a domestic servant named Jacques Renaud. Conversation was in French as none knew much English (Robbins and Howson 1958). All had come to America as representatives of King Louis XVI and the French monarchy. Their mission was to study American plants and to collect plants, seeds, and other useful products of natural history to send back to France (Chinard 1957). Even logs of two to three feet in length were to be shipped to France. Michaux was required to report on the usefulness of each specimen, which would be subjected to empirical tests of strength, durability, and utility (Spongberg 1990). Plants useful for construction and carpentry, for agriculture, and for medicines were of primary interest, while those for ornamental purposes were of secondary concern (Robbins and Howson 1958). Michaux was required to send not only plants to France, but also specimens taken from the North American fauna. These precious products from the New World's bounty of nature would be used to restore France's forests, to enrich the royal gardens and parks, and to enhance the royal hunting grounds, waterways, and forests of Rambouillet, the destination of most of the shipments from North America. L'Abbé Nolin was in charge of cultivating the plants sent to Rambouillet (Savage and Savage 1986). Figure 1, in the previous chapter, shows the primary chain of command of individuals involved in André Michaux's mission to America.

Forces on both sides of the Atlantic favored the mission between France and America. In America, Hector St. John de Crèvecoeur (1735–1813), author of *Letters from an American Farmer* (1782) and later French consul to New York, Connecticut, and New Jersey, wrote to New Jersey's first governor, William Livingston (1723–1790), two years before Michaux's arrival concerning the acquisition of land and establishment of a garden in New Jersey. Crèvecoeur promised the governor full cooperation from Louis XVI, and the New Jersey legislature voted to allow the French to purchase the land. Likewise, French diplomat Louis-Guillaume Otto, Comte de Mosloy (1754–1817), had requested sending a French botanist to America to provide a source of woody plants for French nurseries, forests, and parks (Rey 1954). In a letter of 1784 from l'Abbé Nolin to Madame Helvetius, an intimate friend of Benjamin Franklin, the Abbé indicated an interest in obtaining plants from America (appendix 2-2). Le Monnier urged Louis XVI to send André Michaux to the New World. Le Monnier had already been instrumental in sending Michaux to Southwest Asia, Jean-Baptiste Christophe Fusée Aublet (1720–1778) to Guyana, René Louiche Desfontaines (1750–1833) to North Africa, and Jacques-Julien Houttou de la Billardière (1755–1834) to Lebanon (Rey 1954).

Great anticipation and excitement undoubtedly permeated André Michaux's entire being as he left *Le Courrier*. Visions of finding new plants and other unknowns in the vast, foreign wilderness of North America, "les wilderness," undoubtedly filled the explorer's mind. To be sure, he hoped not only for things that would excite him, as he often remarked in his *Journal*, but also for things that would benefit France and all of mankind. Without doubt his anticipation and excitement were mixed with apprehension—apprehension of traveling in a strange, largely unexplored land, of unseen hardships and dangers that were ahead, and of having to communicate in English, a language he did not know very well.

It was truly a momentous event in the annals of botany when the Michaux party left the ship and stepped onto American soil. Here marked the beginning of countless steps André Michaux, accompanied by his son François André for the first four and one-half years, would take in the New World. His accomplishments during the eleven-year (1785–1796) stay are impressive.

Even though Michaux had received training from some of the very best botanists of the day, it is still amazing that he could recognize so many plants in a foreign land and determine on the spot, very often correctly, that a plant represented a new species or a new genus. The French government could have made no better choice for their official representative than André Michaux.

One of the first persons André Michaux met on American soil was Louis-Guillaume Otto, Comte de Mosloy. Otto had been in the United States for six years and was general consul of France and chargé d'affaires in New York. The count had been unsuccessful in obtaining plants and animals from area landowners to be sent to France. Otto's great satisfaction with Michaux's arrival was beyond expression, as the count revealed in a letter dated November 25, 1785, to Comte d'Angiviller (Robbins and Howson 1958; Savage and Savage 1986). Michaux was required to forward all shipments to Otto in New York for final shipment to France. Captains of vessels had orders from the king to expedite the cargos to ensure their safe arrival to France.

It was not long before Michaux purchased maps of the United States to familiarize himself with the new land and an English dictionary to tackle the language problem, which he successfully overcame (Robbins and Howson 1958). On the fourth day, November 17, after arriving in the New World, Michaux was in New Jersey searching and collecting in Elizabeth-town, now Elizabeth, and on Long Island (Duprat 1957). A few days later he was in New-Ark, now Newark, New Jersey, and Harlem, New York, and then went back to Snake Hill, New Jersey (Robbins and Howson 1958). Despite his arrival late in the season for collecting, when most fruits and leaves were gone and the frigid weather had frozen the ground, by mid-December Michaux had amassed a collection of five boxes of seeds and plants representing the American chestnut, *Castanea dentata* (Marshall) Borkh.; tuliptree, *Liriodendron tulipifera* L.; cranberry, *Vaccinium macrocarpon* Aiton; highbush blueberry, *Vaccinium corymbosum* L.; red and white oaks (*Quercus* spp.); shoots from the sweetgum, *Liquidambar styraciflua* L.; and other plants. These first treasures were soon shipped to France along with eighteen "partridges" or bobwhite quail, *Colinus virginiana* L. This shipment was the first of many to follow. The drive, vigor, and enthusiasm Michaux displayed in quickly assembling this first shipment, under most unfavorable conditions, never diminished.

On March 28, 1786, Michaux purchased land to establish a garden-nursery or *pépinière* where plants could be temporarily held, grown from seeds, and shipped to France with relative ease. The 29.75-acre plot was located between a branch of the Hackensack River on the west and the Hudson River on the east in Bergen County (now Hudson County), New Jersey. New York City was about six miles from the New Jersey pépinière (Robbins and Howson 1958). Pierre Paul Saunier (ca. 1751–1818) was placed in charge of the New Jersey facility, which became known to the local people as the Frenchman's Garden (Robbins and Howson 1958). A detailed description of the layout and plantings of the New Jersey pépinière, based on information obtained from descendants of Saunier's son, Michael (1794–1844), is given by Rusby (1884).

At the end of August 1786, Michaux, François André, and a servant sailed for Charleston (Michaux 1793). By November of that year, another garden-nursery was established near Ten Mile Station on the Southern Railway, about ten miles from Charleston, South Carolina (Coker 1911). During most of Michaux's stay in America he resided at the 111-acre Charleston pépinière. On many of Michaux's specimens the locality is given as "Basse." This refers to Basse Caroline (South Carolina), which most probably includes his garden and nearby areas. Michaux grew many plants in the Charleston garden that he had collected either as individual specimens or as seeds from Florida, the Bahamas, and other parts of America. Michaux also planted seeds in the Charleston garden that had been sent to him from the Cape of Good Hope, China, Mexico, and other foreign countries. Lists of these species constituted most of the *Journal* entries for 1790; Michaux wrote the lists at the end of the notebook for 1795–1796. Sargent (1889) omitted these lists from his transcription of Michaux's *Journal*. The lists of foreign plants can be obtained from the archives of the American Philosophical Society's library.

No trace of either garden exists today (Robbins and Howson 1958; Joyce 1988). Today the site of the New Jersey garden is divided among the Hoboken Cemetery, warehouses, railroad tracks, and marshlands along Cromakill Creek, and Tonnelle Avenue (the Hackensack Turnpike) passes through it (Robbins and Howson 1958). Although attempts were made by several organizations in 1924 to save the triangular-shaped Charleston garden, nothing came of these efforts. As late as the mid-1940s, remnants of Michaux's plantings still existed at the site (Hunt 1947). The property came under ownership of the U.S. Army in World War II. The army cleared the higher ground for watchdog kennels, and in March 1943, army living quarters were constructed in the immediate vicinity of the garden. Then a large army housing development was constructed with a road network and sewage disposal system. Later a large pit was dug for fill dirt for airport runways located nearby. The pit became a city sewage treatment pond. After the war, the city of Charleston regained ownership of the garden, but in March 1961, 5.35 acres were sold to the Air Force for erection of a radio transmitter and receiver station (Joyce 1988).

Saunier never returned to France. He married Margaret "Peggy" Ackerman, of Dutch-American descent. They had two sons, Michael and Abraham (1797–1835), and two daughters, Margaret "Peggy" and Angelick (Robbins and Howson 1958). Saunier maintained and cultivated the New Jersey garden in addition to purchasing his own land, which was located nearby. Although the gravestone of Saunier has not been found, his two sons, their wives, and Margaret Saunier are buried in the English Neighborhood Reformed Church's cemetery in Ridgefield, New Jersey. The church was about five miles from the site of the New Jersey garden, just off the Hackensack Turnpike. Saunier and his family were members of the church, which was organized in 1770 (Robbins and Howson 1958).

During his explorations of "les wilderness," Michaux and his workers shipped more than sixty thousand trees and ninety boxes of seeds to France (Chinard 1957; Robbins and Howson 1958). The actual number of trees may be more than seventy-five thousand, as reported in a document written on August 15, 1800, by Michaux to the minister of the interior (Savage and Savage 1986).

Shipping plants, especially live specimens, was very difficult (appendix 2-6). This was before the development of Wardian cases, small glass houses. It took a minimum of a month and a half for plants to reach Paris from New York. Shipping became more complicated after 1788 when the regular line from New York to l'Orient was shut down (Guillaumin and Chaudun 1957). L'Abbé Nolin wrote several times that the plants had arrived in poor condition. In a letter to Michaux of May 7, 1787, he told the botanist to dip the roots of plants in a mixture of clayey soil and cow manure (Guillaumin and Chaudun 1957). Michaux expressed his irritation at Nolin in a letter to Thouin, dated November 6, 1787 (appendix 2-8).

Michaux sent plants to his colleagues Le Monnier and Jacques Philippe Martin Cels (1740–1806) (figure 1) for their nurseries and gardens. Le Monnier's garden, Montreuil, where Michaux had received training from his mentor, was near Versailles, and Cels's garden, Montrouge, was near Paris. Cels is credited with establishing the national forests and in saving the royal parks and chateaux during the French Revolution (Savage and

Savage 1986). Several plants or seeds sent by Michaux to Cels were later described by Étienne Pierre Ventenat (1757–1808), librarian at the Panthéon and member of the Institut de France (Ventenat 1800–1803). Other people who obtained plants from Michaux's shipments were the King's brother and Chrétien Guillaume de Malesherbes (1721–1794), secretary of state (Guillaumin and Chaudun 1957). Louis Auguste Guillaume Bosc (1759–1828), prolific naturalist and later professeur de culture at the Jardin des Plantes, was a good friend of Michaux from their student days under the tutelage of Antoine-Laurent de Jussieu and André Thouin. Bosc arrived in Charleston in 1796 anticipating collaborating with Michaux. Finding the latter already on his way to France, Bosc stayed on for two years studying and collecting myriads of natural history specimens. Bosc never became French consul in Charleston as professed by many authors (Beale 1978). Serious political problems developed with President Adams

Photo 5. Camellia, *Camellia japonica* L. Introduced to America by André Michaux. This plant with its many varieties graces many southern gardens.

Above: Photo 6. Silktree or mimosa, *Albizia julibrissin* Durazzini. Introduced to America by André Michaux. The silktree is widely planted as an ornamental tree throughout eastern North America.

Left: Photo 7. Crape-myrtle, *Laegerstroemia indica* L. This common landscape shrub of the South was introduced to America by André Michaux.

and France, and Bosc left in 1800 because war was threatening between France and the United States. He brought seeds from the Charleston garden to Paris, some of which were given to Cels for his nursery.

As a measure of goodwill, André Michaux introduced plants in 1785 from the Old World into American soils that today grace lawns and gardens. Among these are the camellia, *Camellia japonica* L. (photo 5); sweet olive, *Osmanthus fragrans* Loureiro; silk tree or mimosa, *Albizia julibrissin* Durazzini (photo 6); crapemyrtle, *Laegerstroemia indica* L. (photo 7); and the pomegranate, *Punica granatum* L. from Persia. Contrary to many authors who have stated that Michaux was first to introduce the ginkgo, *Ginkgo biloba* L., to the New World, the plant came first from England in 1784 to William Hamilton's (1745–1813) estate, The Woodlands. The famed gardens and horticultural showplace with rare plant collections were located in Pennsylvania, not far from the Bartram Garden (Spongberg 1990). Michaux did bring the ginkgo in 1785 to America on behalf of the French government.

In addition to plants, living specimens of whitetail deer, *Odocoileus virginianus* Zimm.; wood duck or summer duck, *Aix sponsa* (L.); and bobwhite quail found their way to France (Chinard 1957; Robbins and Howson 1958; appendix 2-6). Michaux even preserved skins of mammals and birds as well as collecting insects for his beloved country. His specimen of the alligator, *Alligator mississippiensis* Daudin, a hatchling said to be from the shores of the Mississippi River, is the holotype for the species (Daudin 1802; Ross and Ernst 1994). The location is puzzling because, according to the *Journal*, Michaux never traveled near the areas of the Mississippi where the alligator is found today.

The "French Wanderer," as Michaux was often called by many American inhabitants, and his party ventured into mosquito-infested swamps and rivers where they were devoured by pests. Michaux ascended high mountains, even during winter, and once walked twelve miles with a lame horse across snow-covered, frozen ground on his way to Louisville, Kentucky. His toes on the right foot became severely inflamed during this trip. To walk twenty-five to thirty miles or more a day in all kinds of weather was not uncommon for the explorer. It is amazing that André Michaux accomplished his mission at all considering the hardships he endured. His greatest fear was neither death nor disease, but, as expressed by the man himself, "leaving discoveries to be made by those who shall come after me."

The official botanist of King Louis XVI was a patriot even though the French government failed to keep its financial commitment to Michaux for the full length of his North American stay. For his last seven years in America Michaux received no salary nor compensation from the government. His determination to complete the explorations was so compelling that he sold three black men whom he had bought with his own money, and also borrowed against his estate in France to finance the trip into Canada (Savage and Savage 1986). The thirty thousand francs owed to Michaux by the government were never paid despite numerous written pleas.

He apparently rejoiced at the change from a monarchy to a republic that occurred in September 1792. His *Journal* entry for August 30, 1794, reveals that after climbing Grandfather Mountain in North Carolina, which he thought to be the highest mountain in North America, he sang the Marseillaise and yelled out "long live America and the French Republic, long live liberty, etc., etc."

Michaux became acquainted with numerous notables, including Benjamin Franklin; Thomas Jefferson; George Washington; George Rogers Clark, leader in the Revolution, conqueror of the Old Northwest, and elder brother of William Clark of the Lewis and Clark Expedition; William Gerard De Brahm (1717–ca. 1799), surveyor general of British East Florida (1766–1770) and later resident of Philadelphia; Isaac Shelby (1750–1826), first governor of Kentucky; General William Moultrie (1730–1805), governor of South Carolina; William, John Jr., and Moses (1732–1809) Bartram; John Fraser; Dr. Benjamin Smith Barton; David Rittenhouse (1732–1796), astronomer, mathematician, and president of the American Philosophical Society from 1791 until his death; William Hamilton; and Dr. Benjamin Rush, Philadelphia physician and signer of the Declaration of Independence, who treated patients during the yellow fever outbreak of 1793.

Michaux also knew wealthy Carolina planter, Revolutionary patriot, diplomat, and U.S. senator (1789–1795) Ralph Izard (1742–1804) and Major General Thomas Pinckney (1750–1828), attorney, governor of South Carolina, (1784–1789), and minister to England. Pinckney's Treaty (or Treaty of San Lorenzo) with Spain in 1795 provided for free navigation of the Mississippi by Americans and the Spanish. The treaty also set the boundaries of East and West Florida. Thomas's older brother General Cotesworth Pinckney (1746–1825) was honored when Michaux named the fevertree *Pinckneya* (Savage and Savage 1986). Pinckney was captured by the British at Charleston in 1780 and held until 1782. He was one of the leaders at the Constitutional Convention, and between 1789 and 1795, he declined offers to command the U.S. Army, serve on the Supreme Court, and serve as secretary of war and secretary of state. In 1796, however, he accepted the post of minister to France, but the revolutionary regime refused to receive him and he was forced to proceed to the Netherlands. When he returned home in 1798, he found the country preparing for war with France. According to his *Journal*, Michaux visited Pinckney repeatedly when he returned to Paris.

André Michaux and William Bartram

Exactly when André Michaux met William Bartram for the first time is not known because the first part of Michaux's *Journal* was lost during the shipwreck of 1796. In a letter dated June 11, 1786, written from Philadelphia to an unnamed recipient in France, probably Comte d'Angiviller, Michaux said that he saw the *Franklinia*, the only new and interesting tree in the Bartram Garden. On July 15, 1786, Michaux wrote two letters from New York: one addressed to d'Angiviller and one to d'Angiviller's personal secretary, Cuvillier. In the Cuvillier correspondence Michaux mentions that he had returned from a June visit to Pennsylvania, Maryland, and parts of Virginia (appendix 2-4). The lengthy and detailed discourse on Philadelphia and its inhabitants suggests that the June visit was most probably Michaux's first trip to that city even though he had written to Le Monnier on December 9, 1785—six months earlier—that he would leave shortly for Philadelphia (Duprat 1957).

In the letter of July 15 to d'Angiviller, Michaux reiterated the places that he had visited (appendix 2-5). He specifically said that he was well received by Dr. Franklin and that Franklin promised him letters of recommendation for his upcoming explorations to Carolina and Georgia. Furthermore, Michaux wrote to d'Angiviller that General Washington had offered the botanist the privilege of sending his collections to Mount Vernon for deposit before shipping to France. An entry in Washington's diary indicates that Michaux visited him on June 19, 1786: "A Monsr. Andri Michaux—a Botanest sent by the Court of France to America came in a little before dinner with letters of introduction from the Duke de Lauzen and Marqs. de la Fayette" [Marquis de Lafayette (1757–1834)]. After dining with the future President of America, Michaux returned to New York.

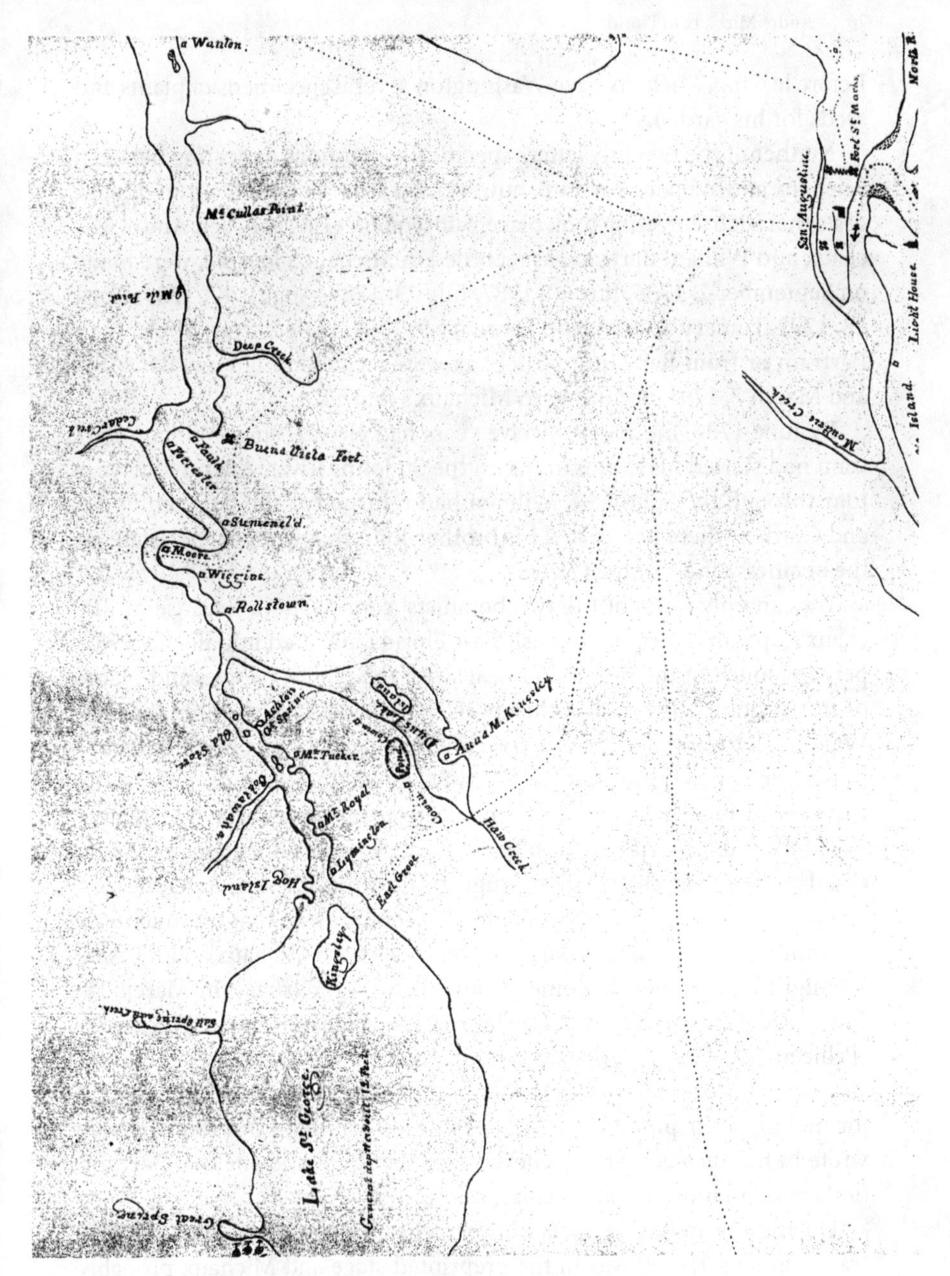

Map 1. Map of East Florida showing sites along the St. Johns River, in the area of Lake George. Note the Indian trails connecting the St. Johns River and St. Augustine. The trail from St. Augustine to the Wiggins's place was called Mount Pleasant and is believed to be the one taken by Michaux. Photostat is from a large map titled "Undated Manuscript Map of the Coast of Florida from Little Cumberland Island to the South Lagoon." The map was enclosed with a letter from Andrew Jackson to John Quincy Adams, October 4, 1821. Artist of the map is unknown. By permission of the St. Augustine Historical Society.

Before leaving, Michaux gave Washington several specimens of plants and seeds for his garden.

Neither of the two July letters specifically mentions a meeting between Michaux and William Bartram; however, the letter of June 11 strongly suggests that such a meeting took place. Subsequent visits between André Michaux and William Bartram, as recorded in Michaux's *Journal*, were made on September 2, 1786, July 30, 1787, July 23, 1789, August 17, 1789, April 28, 1792, January 3, 1794, and February 9, 1794. Letters from Michaux to Bartram or from Bartram to Michaux were written on January 21, 1789, and March 7, 1791, according to Michaux's *Journal*.

By June 1786, more than twelve years had elapsed since William Bartram had started his explorations on the St. Johns River. After a four-year tour through the Southeast, William had returned to the Bartram house and Garden where he assisted his brother John Jr., owner of the estate, in the business affairs of the Garden.

We can only guess at the two botanists' conversation. Because of Michaux's upcoming trip to Spanish East Florida, planned initially for 1787, perhaps much of the discussion dealt with the plants, people, and places of the St. Johns River and other parts of Florida. It is possible that Job Wiggins, Bartram's friend and traveling companion, was mentioned because Michaux, upon departing from St. Augustine, went directly to the Wiggins' residence near Rollestown, located near present-day San Mateo, Putnam County, on the eastern bank of the St. Johns River (map 1). Several days before Michaux left St. Augustine for the St. Johns River, he sent a man to the river to secure a boat for their journey on that thoroughfare.

Another indication that Bartram conversed with Michaux about plants he might find in Florida comes from Michaux's *Journal*. On March 18, 1788, Michaux wrote that he recognized on the banks of the Northwest (Pellicer Creek of today) and Matanzas Rivers an *Andromeda* with leaves shaped like those of an almond and with hollow stalks that Indians used for making their pipes. The plant was not blooming; however, Michaux wrote in his *Journal* that he believed the plant to be the one that Bartram designated to him as *Andromeda formosissima* (photo 8).

During the initial contacts between the two botanists, William Bartram's famous *Travels* was in the preprinted stage and Michaux probably saw an unpublished copy. Johann David Schoepf (1752–1800) saw Bartram's unprinted manuscript on the nations and products of Florida when he visited William in 1783 (Harper 1958). *Travels* was finally published in 1791, about three years after Michaux had left Florida.

Photo 8. Coastal doghobble, *Agarista populifolia* (Lam.) Judd, from Michaux's herbarium. Michaux said this plant was *Andromeda formosissima* of Bartram and that Indians used the plant to make their pipe stems. IDC 57-10. By permission of the Inter Documentation Company, Leiden.

Many of the places visited by André Michaux and William Bartram were the same, but the route taken by Michaux up the St. Johns differed from the one taken by William Bartram in May 1774. Instead of crossing the middle of Lake George after leaving Drayton Island, as Bartram had done, Michaux went along the west coast of the lake and then to the south entrance of the St. Johns. While going along the west coast of Lake George, Michaux entered Salt Springs River and located the bubbling springs and found the yellow anisetree, *Illicium parviflorum* Michx. ex Vent. (photo 9). He was familiar with this plant from a visit to the Bartram Garden. The next day Michaux visited Silver Glen Spring and spent the night there.

Michaux, unlike William Bartram, did not traverse the east coast of

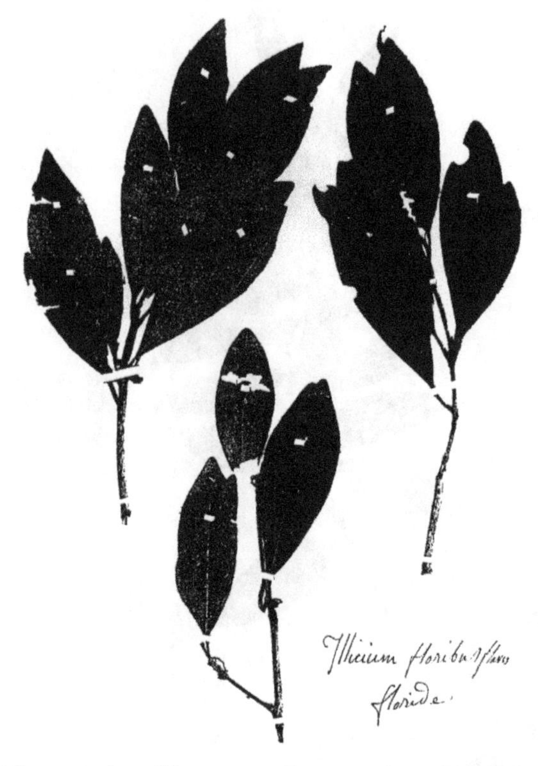

Photo 9. Yellow anisetree, *Illicium parviflorum* Michx. ex Vent., from Michaux's herbarium. IDC 73-3. William Bartram first discovered this plant on January 24, 1766, but André Michaux is responsible for having it named. By permission of the Inter Documentation Company, Leiden.

Lake George. Except for place names such as Lake George and St. Johns River, Michaux did not use Bartram's names for the shell mounds (e.g., Mt. Hope, Mt. Royal), the trading stores (e.g., Spalding's Upper and Lower Stores), and rivers. Instead, he used his own names and descriptions to describe the sites, such as Alligator Point, located opposite the east bank of Lake George. Neither John nor William Bartram used this place name. William Bartram's southernmost point of exploration was at Blue Spring in Volusia County. This is a shorter distance on the river than he and his father had traveled during their trip in 1765–1766. Michaux visited Blue Spring and went eleven miles beyond that site before making the return trip up the river and back along the west coast of Lake George.

In addition to the comparable grounds explored in Florida by both botanists, there are other similarities between the two individuals. Both men had extensive botanical training and both were keen observers in the field. Michaux and Bartram were acquainted with many of the same colleagues. Both kept detailed records of their observations. Bartram's famous account in *Travels* where he describes his experience with alligators at his Battle Lagoon, the present Mud Lake near Lake Dexter in the St. Johns, is similar to that written by Michaux in his *Journal*. Both naturalists wrote a detailed, rather lengthy, anatomical description of the alligator and both said that the reptile's bellow was like that of a bull.

In contrast to these similarities between William and André, there were marked character differences between the two. Michaux refused to get into debt financially, unlike William. This is one reason Michaux declined Jefferson's offer to undertake the proposed expedition to the West that later became the Lewis and Clark expedition in 1803. He handled his finances and other affairs in a more businesslike manner than did William. Michaux was ambitious, aggressive, impatient, terse, and did not display his emotions as did William. Michaux's drive and passion for discovering plants were much stronger than those of William Bartram. At times, Michaux worked as though time were running out. William was more amiable, easy going, somewhat lazy, and, as remarked by Slaughter (1996a), a philosopher of emotions. Michaux's personality resembled that of John Bartram more than that of his son.

Michaux was seven years younger than William Bartram. A very cordial relationship existed between the two botanists. While at the Bartram Garden on January 3, 1794, Michaux asked William for a list of plants that he desired from the Frenchman's upcoming trip to the Carolinas. On another occasion, Michaux brought William Bartram plants of the twinleaf, *Jeffersonia diphylla* (L.) Persoon. Michaux had discovered the plant in the mountains of Virginia.

Not only did Michaux have a good relationship with William, but he also knew William's brothers, Moses and John Jr. Moses, William's older brother, who with his brother Isaac (b. 1725) prospered in the apothecary business, visited the South Carolina pépinière in the spring of 1791. He gave Michaux valuable seeds that had been collected in China (Savage and Savage 1986).

Data gathered on the flora of Florida in the 1700s are limited mainly to the observations, collections, and writings of Bernard Romans (1720–ca. 1784), John and William Bartram, and André Michaux. Of these, the Bartrams and Michaux took trips specifically to locate and document plants. It is evident from the records of these naturalists that not all plants observed were recorded. Michaux, for example, did not record in his *Journal* water lettuce, *Pistia stratiotes* L., growing in the St. Johns River; however, the plant is listed in the *Flora*. We also know, from a handwritten, loose sheet located in an archival box of the manuscript division of the American Philosophical Society, that he saw the lettuce on the St. Johns. The buttonwood, *Conocarpus erectus* L., rougeplant, *Rivina humilis* L., and gray nicker, *Caesalpinia bonduc* (L.) Roxb., and other plants are not in the *Flora*, but are listed either in a seed shipment or in the *Journal*. The yellow canna, *Canna flaccida* L., is not in the Flora, IDC collection, or the *Journal*, but the species is mentioned in a paper written in 1792 by Michaux (Rehder 1923).

Michaux recorded a number of plants that are not found or are uncommon in Florida today, or whose present-day range does not extend to the geographic areas traveled by the French botanist. Among these plants are the pond apple, *Annona glabra* L., *Hypericum denticulatum* Walt., the pawpaw, *Asiminia triloba* (L.) Dunal, hazelnut, *Corylus americana* Walt., American chestnut, *Castanea dentata* (Marshall) Borkb., and possibly the manchineel (*Hippomane mancinella* L.). Some insight into the diversity of east-central Florida's flora in the late 1700s can be gained by analyzing lists of plants found by the Bartrams and Michaux (appendix 1-1). The plants listed for the former were taken from John Bartram's published diary, annotated by Harper (1942), and from William's *Travels* and the Fothergill report also annotated by Harper (1943, 1958).

John and William's trip up the St. Johns River occurred in November of 1765 and into January and February of 1766. Their assigned task was to locate the river's source and to determine the economic value of the region and how well it could be exploited (Slaughter 1996a). There was renewed talk and interest in building a canal across the state beginning at St. Augustine (Hetrick 2000). In 1774 William explored the St. Johns, making more than one trip, and extended his explorations to the west coast of Florida. These trips occurred from April into May as well as in August–September. Florida was under British rule at the time the Bartrams made their trips. Michaux visited Spanish East Florida in February through May of 1788. Michaux's travels extended along the east coast of Florida and up the St. Johns River. The French botanist apparently did not explore the Florida scrub visited by William Bartram on the west side of Lake George in Marion County. Although the Bartrams spent more time in Florida than did

Michaux, the latter explored more of the east coast than did John or William.

At least 235 species of plants were found in Florida by the Bartrams (appendix 1-1). Of these, 52 species were also documented by André Michaux. There are 73 species of plants that Michaux found in Florida that neither Bartram recorded. Many of these occurred along the east Florida coast. Michaux reported more species of grasses, sedges, lichens, and ferns than did either Bartram.

Most species documented by the Bartrams and Michaux are species that can still be found in Florida. Both species of silverbells, *Halesia carolina* L. and *H. diptera* J. Ellis, and the sourwood, *Oxydendrum arboreum* (L.) De Candolle, apparently are less common today in central Florida than they were at the times of the Bartrams and Michaux. None of them reported pitcherplants from central Florida, though several species were found in other states and William Bartram reported "*Saracinia lacunosa*" from Pensacola. This name applies to the hooded pitcherplant, *Sarracenia minor* Walter, which does not occur in the Pensacola area (Harper 1958).

Rare plants that were found by the Bartrams and Michaux are as follows: Bartram's ixia, *Calydorea caelestina* (W. Bartr.) Goldblatt & Henrich; yellow anisetree, *Illicium parviflorum* Michx. ex Vent.; and Okeechobee gourd, *Curcubita okeechobeensis* (Small) L. H. Bailey. All species, except for the ixia, still occur on the St. Johns River. The ixia has not been seen near the river area for many years. Harper (1943, 1958) concluded that the species was located by William in the area of today's Lake Dexter. No locality accompanies the specimen collected by Michaux. Both the anisetree and gourd occur today on the St. Johns River. One plant described by William Bartram that has evoked much controversy is the Florida royal palm, *Roystonea regia* (Kunth) O. F. Cook. He described this plant in his *Travels* as occurring near today's Manhatten, Lake County, Florida (Harper 1943, 1958). Michaux did not mention seeing the stately palm even though he passed the area where William Bartram saw the plants in 1774. Nor was it documented by John Bartram during the trip up the St. Johns River in 1765–1766.

3

The Two Floridas

Although André Michaux arrived in East Spanish Florida during the Second Spanish Period, the botanist wrote accounts in his *Journal* pertaining to people, places, and events that occurred during the earlier twenty-year occupancy of Florida by England. To fully appreciate Michaux's observations recorded in his *Journal* while exploring Florida, a digression here is appropriate to recount the history of both British Florida and Spanish Florida.

British Florida: 1763–1783

Spain, concerned with possible English domination of America, made a grave mistake in December 1761 by siding with France and declaring war on Great Britain in the Seven Years' War (1754–1763), or French and Indian War of American history. In August of 1762 the British had captured Spain's gateway to the Caribbean—Cuba and its thriving port of Havana. Havana was the pride and center of Spanish America, the bulwark of the West Indies, and the center of the slave trade (Panagopoulos 1966; Corse 1967).

With the unexpected victory by Britain over the two European powers in 1762 and signing of the final peace treaty on February 10, 1763, Spain lost all of its North American territories east of the Mississippi River. To regain Havana, Charles III (1716–1788) of Spain sacrificed Florida to Great Britain, including Fort St. Augustine and the Bay of Pensacola (Mowat 1943). France's losses were disastrous. Louis XV ceded Canada, Mobile, and all of France's southern territories east of the Mississippi to the British, except the town and island of New Orleans (Mowat 1943; Fabel 1996). In an agreement separate from the treaty, the French king gave Louisiana and all French territory west of the Mississippi to Spain (Mowat 1943; Gannon 1993; Fabel 1996). These losses were so severe that France never again became a serious threat to Britain in North America.

Except for 1564 to 1565 when René Goulaine de Laudonnière established a small French colony at Fort Caroline near the mouth of the St. Johns River (Rivière de May to the French), Spanish ownership of "The Flowery Land" had existed for 250 years, beginning in 1513 when Juan Ponce de León (ca. 1460–1521) claimed "La Florida" for Spain. Despite their long occupancy, the Spanish did little toward the development of the land.

With the British takeover of Florida in 1763, the very long First Spanish Period (1565–1763) of Florida history was over. Nearly the entire population of 3,046 Spanish people left Florida, most going to Cuba (Gannon 1993). The last few surviving Timucuan Indians also left Florida when the Spanish withdrew (Sturtevant 1991). When Juan Ponce de León had sailed the coastline in 1513, there were 150,000 Timucua speakers in Florida (Milanich 1996), but after being exposed to a series of smallpox epidemics from European contact, the population was greatly diminished (Hann 1996).

The two decades (1763–1783) that followed under British rule were among the most colorful in Florida's history. Britain did more to develop Florida's land and resources in the twenty years of occupancy than Spain had done in more than two hundred years (Gannon 1993). Slavery was more prevalent during the British domination than it had been during the Spanish occupation (Siebert 1931).

St. Augustine in 1763 had about three hundred houses. The departing Spanish put the torch to a number of buildings in the town. Even the Spanish governor, according to Corse (1967), destroyed his elaborate garden as a protest against the incoming British governor James Grant (1720–1806). Destructive activities were continued even by the British. John Bartram, after returning from St. Augustine in 1766, lamented to his friend Peter Collinson in a letter dated August 26, 1766, on the "very ruinous condition" of Augustine. Bartram wrote that the British soldiers had pulled down half the town to burn for timber. Cultivated gardens were grown over with weeds, orange and fig trees were cut down for firewood, and the English did not make as much use of the abundant sour oranges as did the Spanish (Berkeley and Berkeley 1982). Bernard Romans, in his book *A Concise Natural History of East and West Florida* (1775), quoted a friend, in a letter dated May 27, 1774, who said that "this town is now truly become a heap of ruins." Despite these destructive acts and the addition of chimneys, fireplaces, and sash windows to the houses by the British, the Spanish town changed very little during the twenty-year British Period (Mowat 1943).

The vast area of Florida, at the time the British took possession of the land, extended to the Mississippi River. The huge territory was almost immediately divided by Britain into two colonies: East Florida (capital at St. Augustine) and West Florida (capital at Pensacola) that included parts of present-day Alabama, Mississippi, and Louisiana. The Apalachicola River was the dividing line, established by British surveyor and cartographer William Gerard De Brahm (Tanner 1963; Gannon 1993; Derr 1998). De Brahm was appointed surveyor general for the southern district of North America. His assistant was Bernard Romans, a native of Holland and appointed deputy surveyor in 1770 (Siebert 1929). The northern boundary of East Florida was the St. Marys River. For West Florida, the northern boundary was initially set at 31 degrees of latitude, but was moved to 32 degrees, 28 minutes (Fabel 1996). West Florida's area was larger than that of East Florida.

The greater part of Florida was occupied by Indians. East Florida, for all practical purposes to the British, and later to the returning Spanish, was the northeast section established by the first congress, held at Fort Picolata on November 15–18, 1765, between the Indians and the British. The Indians agreed to surrender the entire east coast of Florida and the region of the St. Johns River and some land west of the river (Sastre 1995). The remaining area of the peninsula was either unoccupied or territory of the Seminoles—Lower Creeks who had moved into Florida and had ceased their connection, in the eighteenth century, with the larger Creek Confederacy of the Southeast. The Seminoles comprised the dominant group of Indians in Florida at the time of Michaux's visit. Earlier Florida groups that were extinct or nearly so included the Ais, located from Cape Canaveral south to Fort Pierce, Jeaga along the southeast coast, Tequesta of the Miami area, Calusa along the southwest coast, Jororo of the south-central peninsula, Mayaca of the St. Johns River area in the central peninsula, eastern Timucua speakers of the northern third of the peninsula from the St. Johns River drainage to the Atlantic Coast and into southeast Georgia, western Timucua speakers across north Florida to the west in the panhandle and into south-central Georgia, and the Apalachee in the Tallahassee Hills region of the eastern panhandle (Milanich 1995; Milanich 1996).

James Grant, the first British governor (1764–1771), a native of Ballindalloch, Scotland, arrived in St. Augustine from England, in August 1764 (Mowat 1943; Gannon 1993). Grant was a seasoned British soldier and officer who had fought in the campaign in Flanders (1747–1748), in Ireland, and in the French and Indian War, when he was taken prisoner by the French at Fort Duquesne. He also participated in conflicts against the Carolina Mountaineers and the Cherokees in 1760 and 1761, but his major role was in the siege of Havana. These activities contributed to his appointment as governor (Mowat 1943; Panagopoulos 1966; Corse 1967).

Grant took office in the residence of the former Spanish governor, Melchor Feliú. Grant's hospitable and genial behavior led to his success as a capable and likable administrator (Corse 1967). Grant occupied the governorship until the summer of 1771 when he returned to Scotland. At that time about three thousand inhabitants, exclusive of the Indians, lived in East Florida; two thousand were in St. Augustine. Grant, on the eve of his departure, wrote that East Florida had a population of 288 whites and 900 black slaves (Fabel 1996).

In 1775, Grant returned to North America and was commander in the battles of Long Island, Brandywine, and Germantown of the Revolutionary War, under Howe. He returned later to Great Britain and in 1791 was appointed governor of Sterling Castle. Five years later, Grant was commissioned to the rank of general (Siebert 1929).

One of the first items on Grant's agenda, as governor, was to increase the population of East Florida. To this end, Great Britain made available generous land grants to individuals who applied and who qualified (Siebert 1931; Mowat 1943). Grant envisioned a feudal-type system where landowners would establish plantations worked by black slaves with whites overseeing the operations. Thus began the establishment of extensive plantations in Florida, mostly occupying the fertile lands along both banks of the St. Johns River and along the East Florida coast (maps 1–7).

By February 1765, about six months after his arrival in St. Augustine, Grant's slaves had cleared land on his own plantation, Villa, located north of St. Augustine. Grant's Villa quickly gained a reputation for producing the most profitable crops of indigo in the colony (Schafer 1982). The excellent dye obtained from the imported indigo plant was in high demand. In 1774, three years after Grant retired from the governorship, his plantation

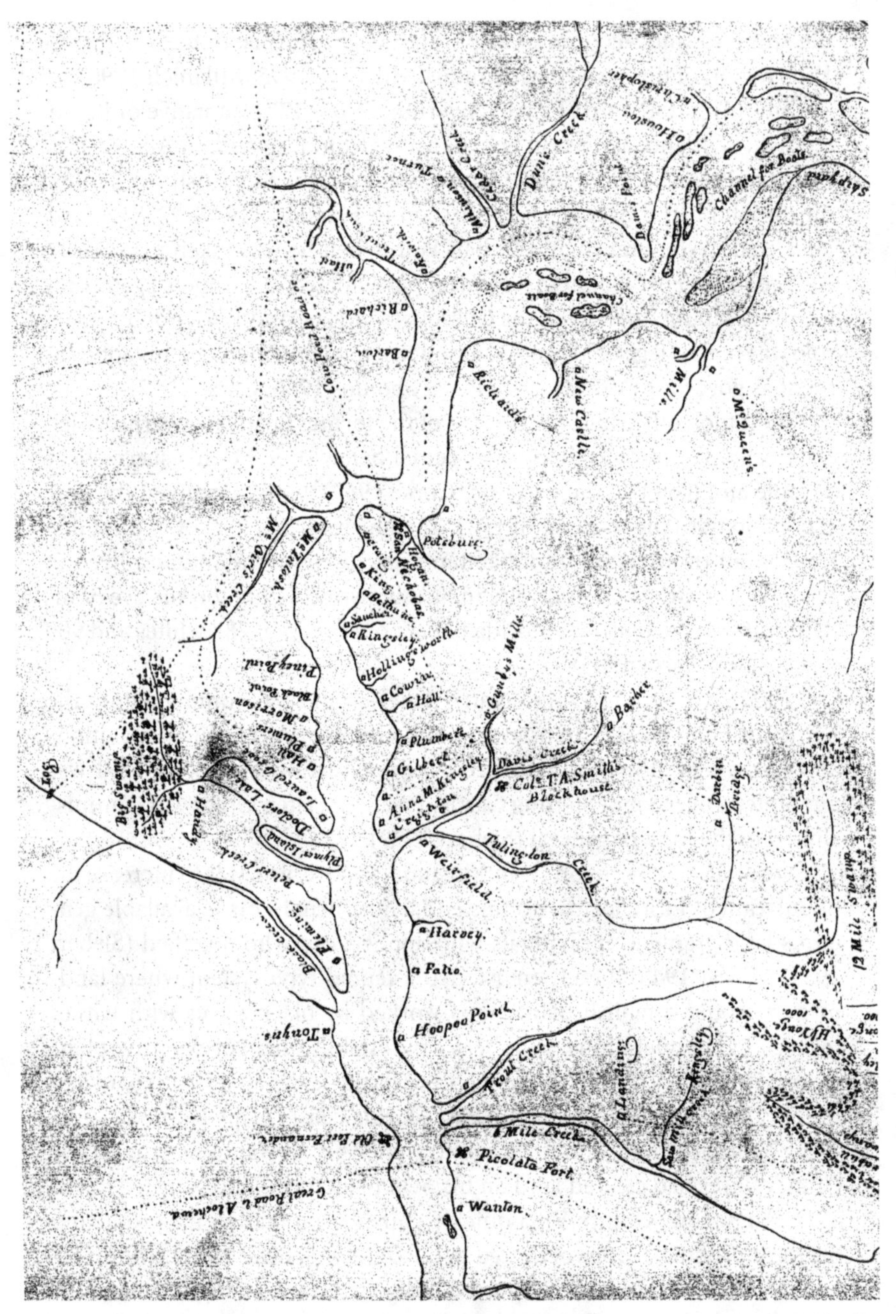

Map 2. Northern section of Map 1, showing locations of Fort Picolata and other sites along the St. Johns River. By permission of the St. Augustine Historical Society.

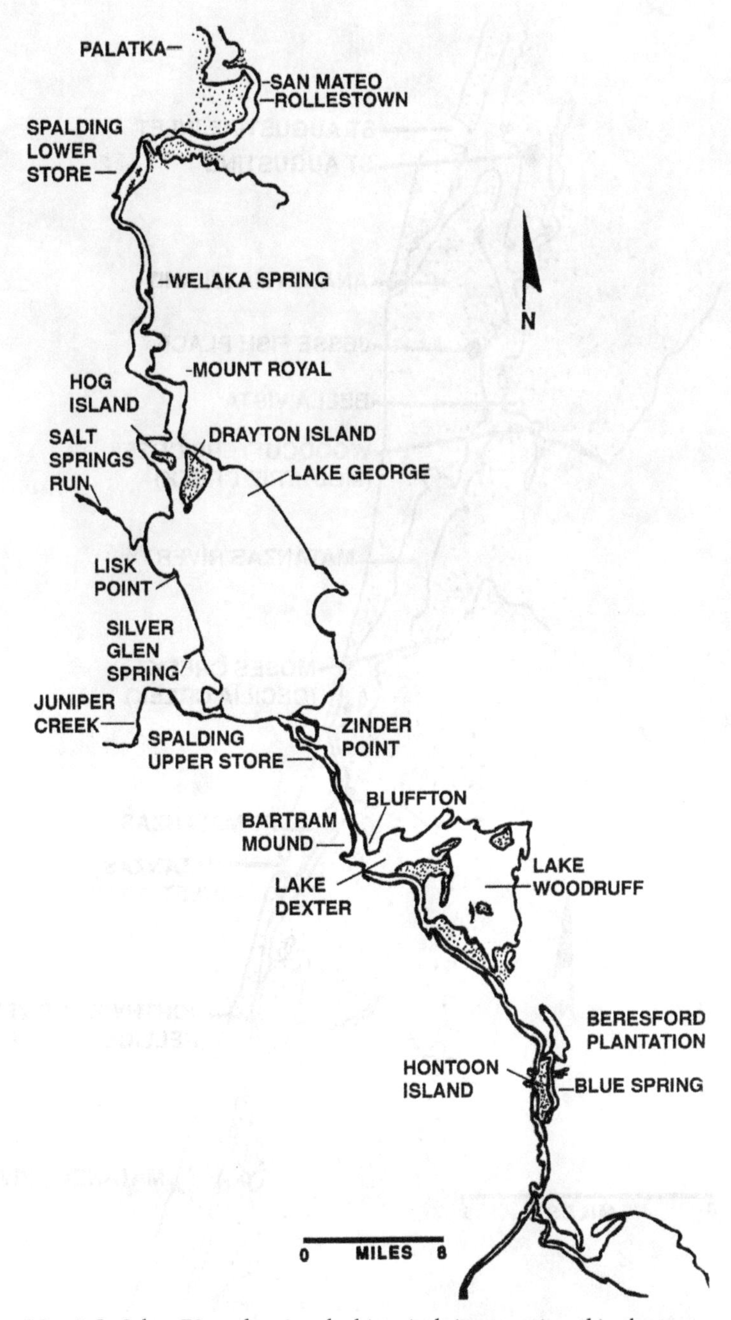

Map 3. St. Johns River showing the historical sites mentioned in the text.

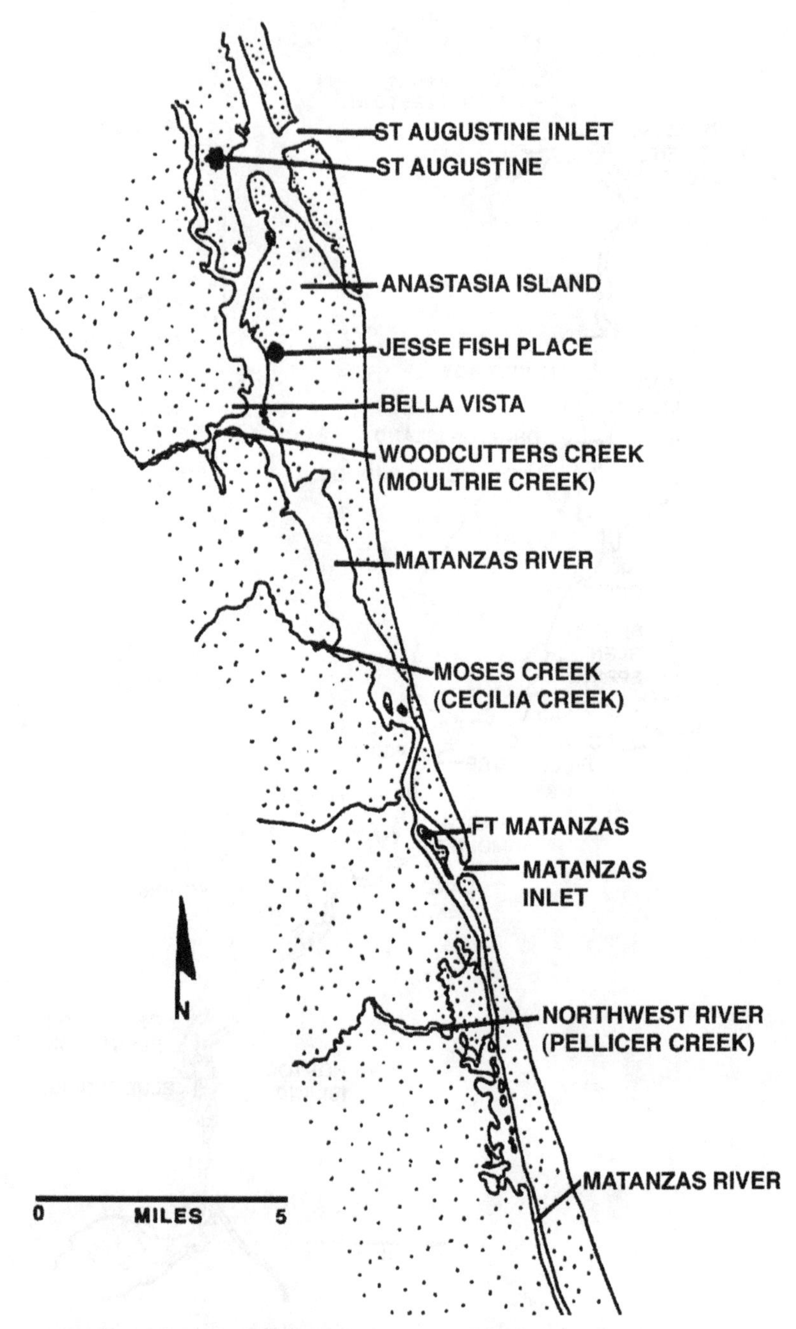

Map 4. Northern section of Florida's east coast at St. Augustine, St. Johns County, Florida.

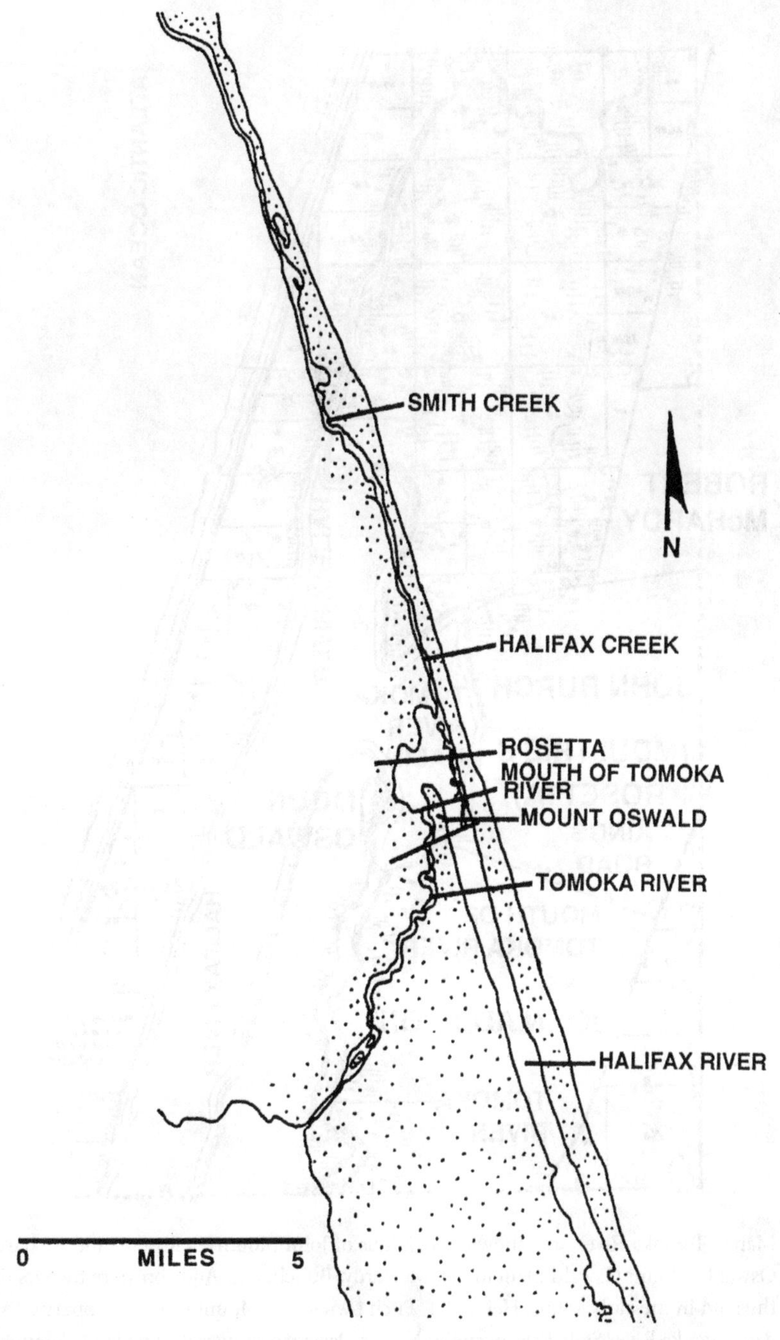

Map 5. Midsection of Florida's east coast at the mouth of the Tomoka River, Volusia County, Florida.

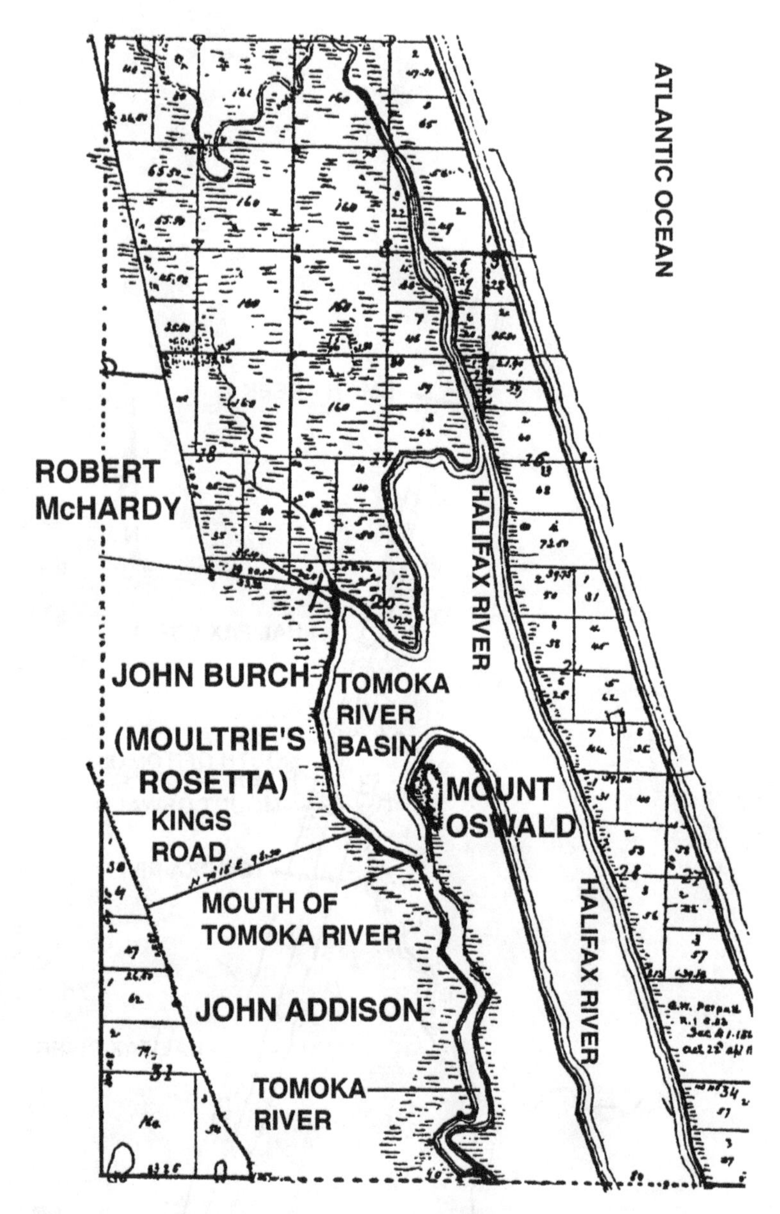

Map 6. Tomoka River area showing locations of John Moultrie's Rosetta and Richard Oswald's Mount Oswald plantations. McHardy, Bunch, and Addison were owners of the land in the early 1800s. Heirs of T. Fitch owned the Mount Oswald property. Today, Tomoka River State Park occupies Mount Oswald and surrounding land. Adapted from a U.S. Survey Map dated 1845.

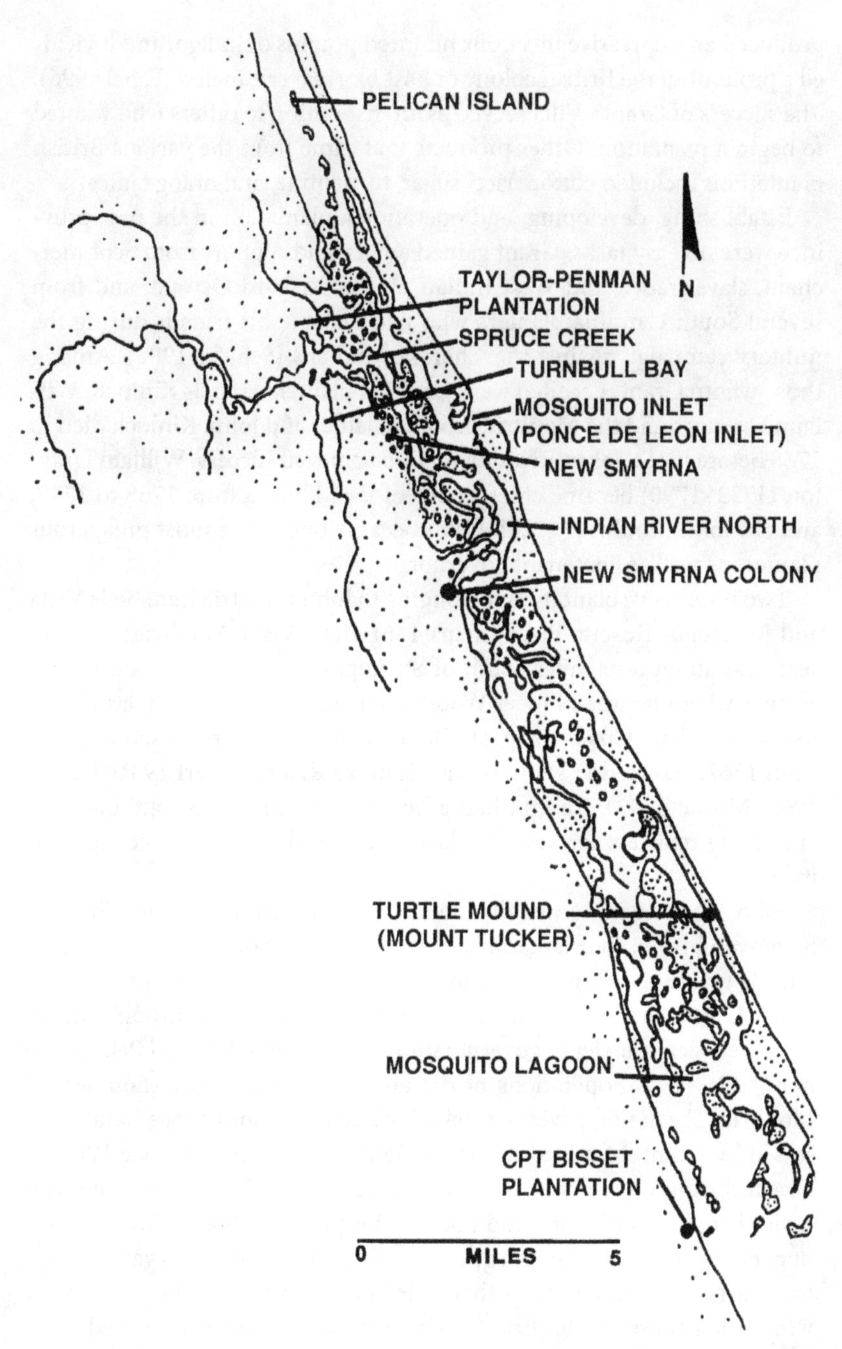

Map 7. Lower section of Florida's East Coast at Mosquito Inlet (now Ponce de León Inlet), Volusia County, Florida.

produced an impressive fifty-four hundred pounds of indigo, and it yielded a profit until the British colony of East Florida terminated (Fabel 1996). The success of Grant's Villa served as an inspiration to others who wanted to begin a plantation. Other products that came from the various British plantations included cotton, rice, sugar, turpentine, and orange juice.

Establishing, developing, and operating a plantation in the new province were no easy tasks. Grant gained advice and support from Scot merchant, slave trader, and West Indian planter Richard Oswald, and from several South Carolina planters who had become his friends during the military campaign against the Cherokee Indians (Schafer 1982). Among those whom Grant recruited were Carolina planters Francis Kinloch, William Drayton, and the Moultrie brothers, James and John. Kinloch died in 1768 before his St. Johns River plantation achieved success. William Drayton (1733–1790) became chief justice of East Florida from 1768 to 1777, and Dr. John Moultrie (1728?-1798) became one of the most prosperous planters as well as lieutenant governor.

Two impressive plantations belonging to John Moultrie were Bella Vista and Rosetta or Rosetta Place (maps 4–6). Bella Vista, Moultrie's country seat, was about four miles south of St. Augustine, on the Matanzas River near where its waters meet Woodcutters Creek or St. Nicholas Creek, today's Moultrie Creek (map 4). The two-thousand-acre Rosetta, dating from 1767, was farther south on the Tomoka River (Siebert 1931; Tanner 1963; Mowat 1943). Rosetta had a house with ten rooms, outbuildings, and living quarters for seventy slaves who worked to produce rice and indigo.

John Moultrie was a wealthy medical doctor from Edinburgh. In 1767, he moved his family, plantations, and slaves from South Carolina to East Florida where he became president of the council and lieutenant governor under Governor Grant. During the interim of July 1771 through March 1774, between transfer of governorship from Grant to Colonel Patrick Tonyn (1725–1804), operations of the goverment fell on the shoulders of Moultrie. The acting governor lacked the charisma and congeniality possessed by Grant. Moultrie's poor relationship with Chief Justice William Drayton, council member Dr. Andrew Turnbull (1719–1792), merchant James Penman, and others did not help his political career. Governor Tonyn, the third and last British governor, sided with Moultrie against Drayton, Turnbull, Penman, and others; his ill feelings toward these men were worse than those of Moultrie. Tonyn, aged forty-nine, had arrived in St. Augustine on March 1, 1774, to assume the governorship. The officers of the garrison and other prominent men in East Florida preferred Andrew Turnbull as their candidate for governor (Siebert 1929).

André Michaux never met John Moultrie, but he did meet his older brother, General William Moultrie (1730–1805) of Charleston, in September 1786. General Moultrie was twice governor of South Carolina and supported the Americans during the Revolution (Mowat 1943). John, on the other hand, gave his support to the British cause.

Many of these British plantations were rather grandiose, as can be illustrated by John Moultrie's Bella Vista, his place of residence. This thousand-acre plantation had a two-story stone mansion of ten rooms with elaborate interiors and arches, a dining parlor, a bowling green, formal gardens, a kitchen garden, walks and drives, parklike areas, fish ponds, and many fruit trees such as dates, oranges, lemons, limes, citrons, figs, and pomegranate (Siebert 1929; Hanna and Hanna 1950). Corn, peas, potatoes, and rice were grown at Bella Vista. A fifteen-hundred-acre tract, considered by Moultrie as an appendage of Bella Vista and obtained later as a grant from Governor Tonyn, was located on Woodcutters Creek. The site contained extensive acreage of yellow, or longleaf pine, and cypress from which turpentine, tar, and lumber were produced (Siebert 1929). The large mansion at Bella Vista was burned by Indians not long after Moultrie and others had departed Florida at the end of the British Period (Gannon 1993).

Opposite Moultrie's Rosetta, and situated in the angle between the Halifax and Tomoka Rivers, was Mount Oswald, dating from 1765 (maps 5 and 6). This plantation was one of four tracts comprising a total of twenty thousand acres granted to wealthy Richard Oswald (1705–1784), who was the province's chief importer of slaves, mostly from Africa (Forbes 1821; Watson 1927; Siebert 1931; Fabel 1996). Oswald was one of the earliest individuals to offer advice to Governor Grant, and to stimulate interest among other potential landowners. Oswald never visited Mount Oswald, but did send 240 slaves to operate the plantation. Here cotton, indigo, and some sugarcane were grown. The Timucuan village of Nocoroco, depicted on a 1605 map drawn by a young Spanish soldier named Alvaro Mexia, had existed on the site of Mount Oswald. Oswald had another twenty-thousand-acre plantation, called Ramsay Bay after his wife, Mary Ramsay, which was never developed. Fearing depredation by the Spanish, Oswald abandoned Mount Oswald in 1780 (Siebert 1929). He was later to serve as Britain's peace commissioner, along with Franklin, Jay, and Adams, during the Treaty of Paris meetings in 1782.

Near the mouth of Spruce Creek (also called Spruce Pine Creek in the

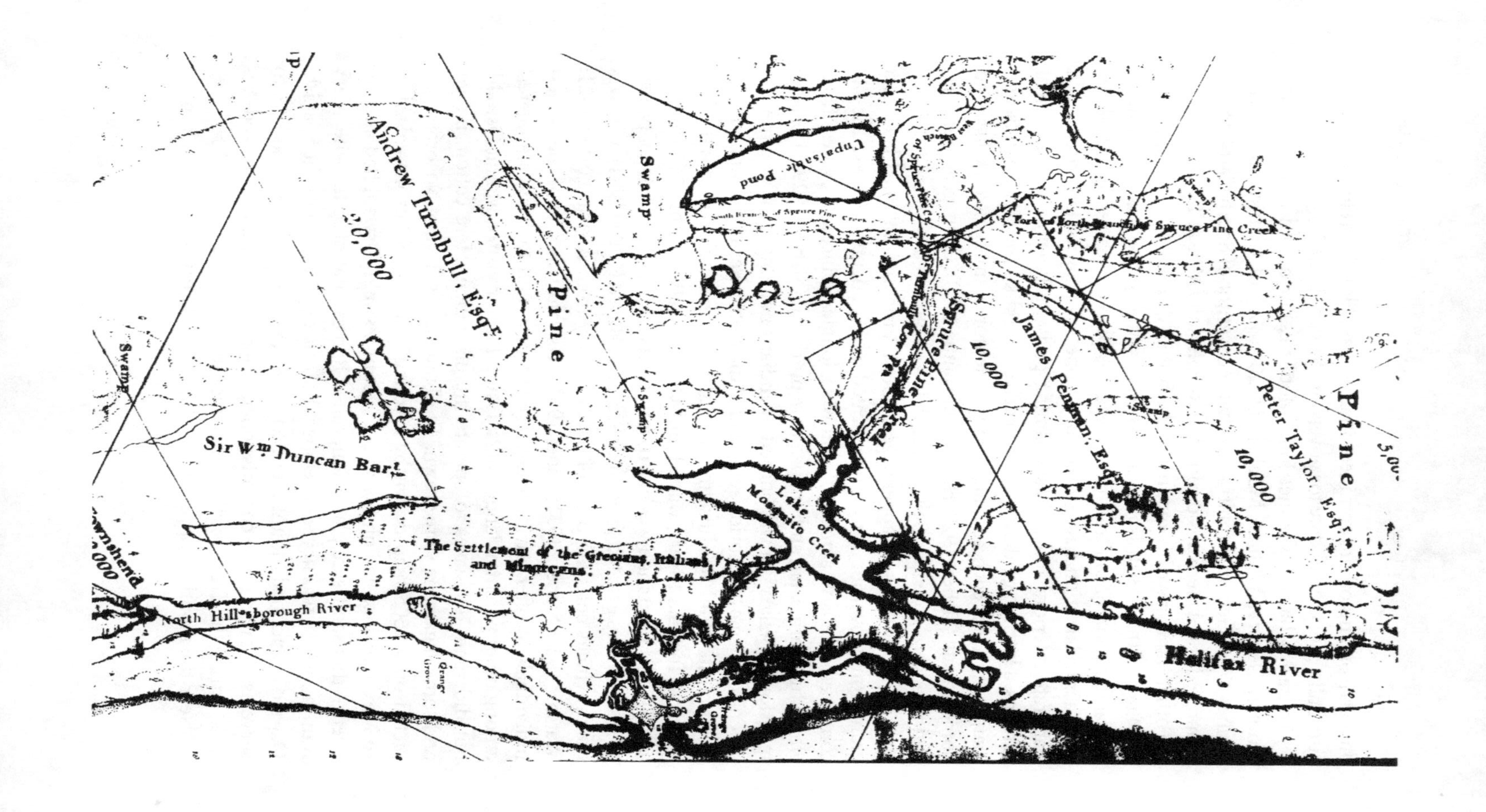

Pine
Peter Taylor, Esq.r
10,000
James Penman, Esq.r
10,000
Halifax River
Spruce Pine Creek
Unpassable Pond
Lake of Mosquito Creek
Orange Grove
Swamp
Swamp
Pine
The Settlement of the Greeks, Italians and Minorcans.
Orange Grove
Andrew Turnbull, Esq.r
20,000
Sir Wm. Duncan Bar.t
North Hillsborough River
Swamp

Mosquito District) was one of the first settlements of this area (maps 7 and 8). Michaux referred to this land, near Spruce Creek, in his *Journal* as the abandoned habitation that belonged to Mr. Penman. A 1769 map of William Gerard De Brahm shows the James Penman ten-thousand-acre grant next to that of Peter Taylor, but D. Schafer (pers. comm., June 2000) stated that Peter Taylor, an absentee owner, had a ten-thousand-acre grant and that James Penman was resident manager and partner in the enterprise. Later the partnership was dissolved and the land divided, with Taylor retaining most of the acreage. Indigo was the major product grown on this plantation (Romans 1775). Penman was one of two prominent merchants of St. Augustine during the Revolutionary Period. He owned several houses and lots in St. Augustine as well as other plantations, including the successful Jericho Plantation on the St. Johns River. Because of personal conflicts with Governor Tonyn, Penman and friends were refused additional land grants. Tonyn unjustly charged Penman, Turnbull, Chief Justice William Drayton, and others as belonging to a faction working against the administration and being disloyal (Siebert 1929).

Farther south on Mosquito Lagoon, then called the North Hillsborough River or South Mosquito River, and fourteen or fifteen miles south of Mosquito Inlet (Ponce de León Inlet of today), stood the three-hundred-acre plantation Mount Plenty or Palmerina owned by Captain Robert Bisset (Watson 1927; Siebert 1929, 1931; maps 7, 9, and 10). Captain Bisset (also spelled Bissett) was one of the earliest settlers of East Florida during the British Period, arriving in 1767 when the area was still richly covered with woods (Siebert 1929).

Bisset amassed nine different tracts of land that totaled nearly ten thousand acres. Mount Plenty was one of the earliest plantations to be established and the most famous of Bisset's tracts. He received the three-hundred-acre land grant in 1768 from Governor Grant but did not settle on Mount Plenty until 1777. The plantation fronted on Mosquito Lagoon and was bounded on the north by Clotworthy Upton's (also spelled Lupton) tract of land (map 10). A few miles south of Mount Plenty was Peter Elliot's sugar plantation (Siebert 1929).

Facing page: Map 8. Land grants and settlement of Turnbull's New Smyrna in the Mosquito District. Adapted from a large sheet of the map titled "A Plan of Part of the Coast of East Florida including St. John's River from an actual Survey by Wm. Gerard De Brahm, Esq. Surveyor General of the Southern District of North America. 1769, done by John Lewis and Samuel Lewis." Map furnished by Georgia Archives and History, Atlanta. By permission of the British Library (Kings MSS 211.12).

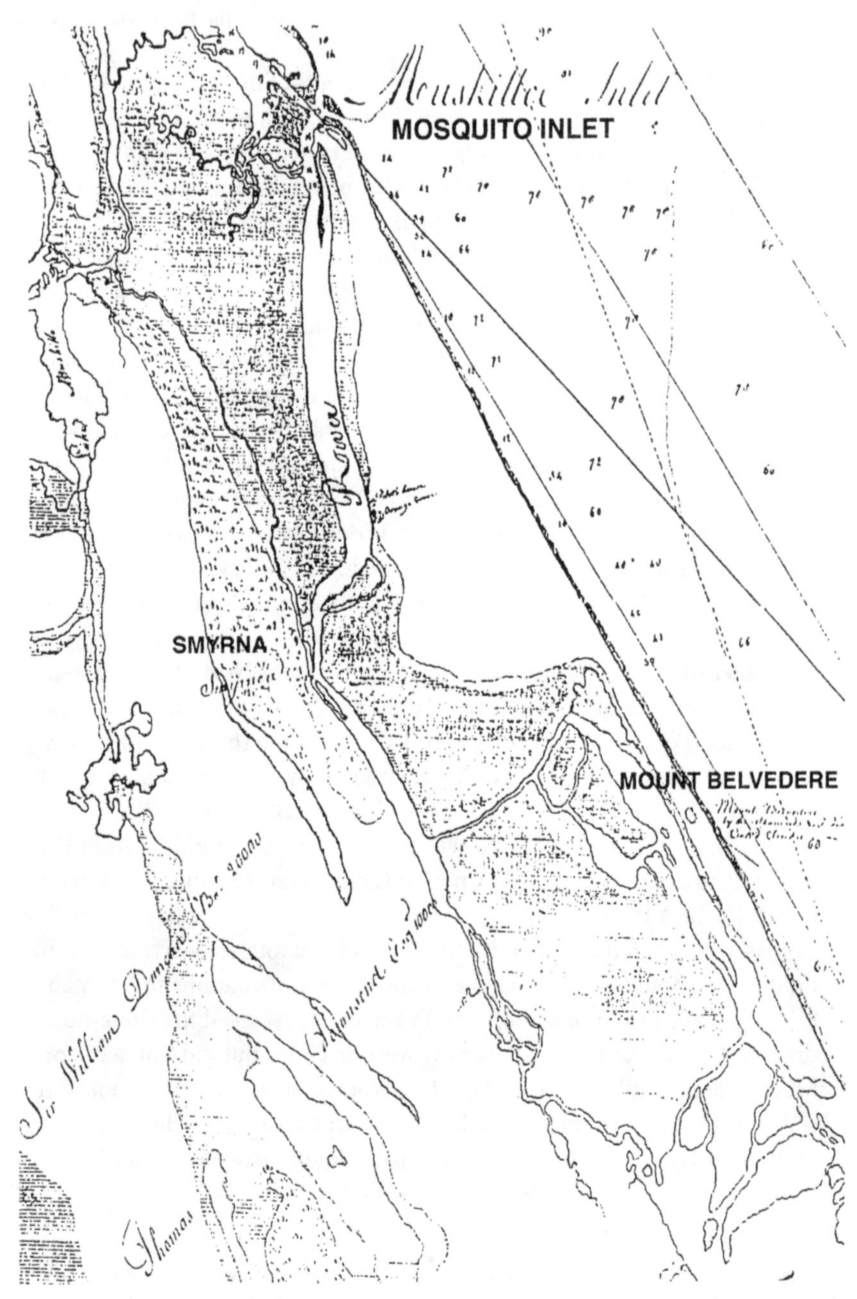

Map 9. Mosquito Inlet. Smyrna is the Turnbull settlement and Mount Belvedere is present-day Turtle Mound, Volusia County, Florida. Adapted from the map titled "East Florida, East of the 82nd Degree of Longitude From the Right Honorable the Lords of Trade & Plantation. Surveyed by William Gerard De Brahm, Surveyor General for the Southern District of North America." Date of map ca. 1771. Original (C.O. 700/Florida 3) in Public Record Office, London.

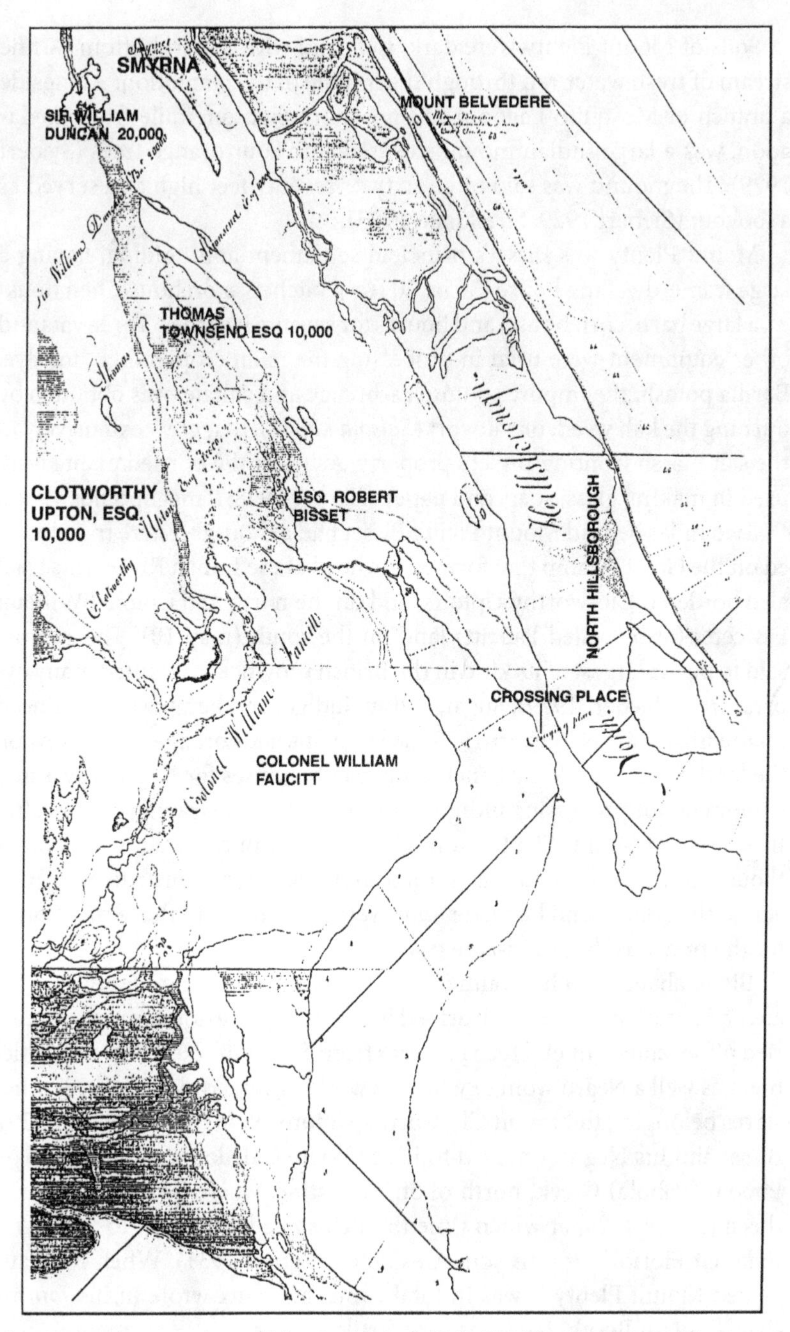

Map 10. Southern section of Map 9. Original (C.O. 700/Florida 3) in Public Record Office, London.

Soils of Mount Plenty were dark, rich, and with a marl bottom. A fine stream of fresh water ran through the middle of the plantation. Alongside a branch of Mosquito Lagoon, and not more than one mile from the lagoon, was a large Indian mound covered with sour orange trees (Siebert 1929). The mound was said to be sixty or seventy feet high and served as a lookout (Siebert 1929; Strickland 1965).

Mount Plenty was Bisset's principal settlement and holding, having a large frame-dwelling house (30 by 20 feet), kitchen, storehouse, hen houses, a large barn, corn house, and houses for seventy Negroes. Three vats and other equipment were used in processing the main crop, indigo, for dye. Barilla potash, the impure sodium carbonate and sulfate, was obtained by burning the kali weed, or saltwort (*Salsola kali* L.), that grew extensively in the salt marsh fronting Bisset's property. Ashes of the burned plants were used in making glass, soap, and paper (Harper 1958; Panagopoulos 1966).

Two miles behind Mount Plenty Bisset had a thousand-acre tract located on the Great Swamp that formed the head of the Indian River. This tract also bordered Clotworthy Upton's land on the north and Colonel William Fawcett's (also spelled Faucitt) land on the south (map 10). The swamp, said to be the largest of its kind in the British province, contained many cypress trees (Siebert 1929). Sugar, cotton, indigo, and hemp were produced on the higher areas (Siebert 1929). A large fine sour orange grove grew on the land. Several buildings, including twenty houses for Negro slaves and equipment for processing indigo, were erected on the tract two years after Bisset settled Mount Plenty. Adjoining the back of this tract was another thousand-acre tract that was settled in 1776. Bisset constructed a road across the swamp and he made improvements similar to those mentioned for the previous thousand-acre tract.

Bisset abandoned his southern lands, including Mount Plenty, in 1779 after a Spanish privateer with armed boats entered Mosquito Inlet and carried off seventeen or eighteen Negroes from Turnbull's New Smyrna settlement as well a Negro woman who was working on naval stores with other slaves belonging to him on Clotworthy Upton's plantation (Siebert 1929). Bisset and his Negroes moved to his 115-acre Caledonia tract located on Pobolo (Pabola) Creek, north of St. Augustine, where he remained until the spring of 1784, at which time the British ceded Florida to Spain. He then left Florida with his son, Alexander (Siebert 1931). When Michaux visited Mount Plenty it was in total ruins. Michaux wrote in his *Journal* that "Captain Besy's" land was very fertile.

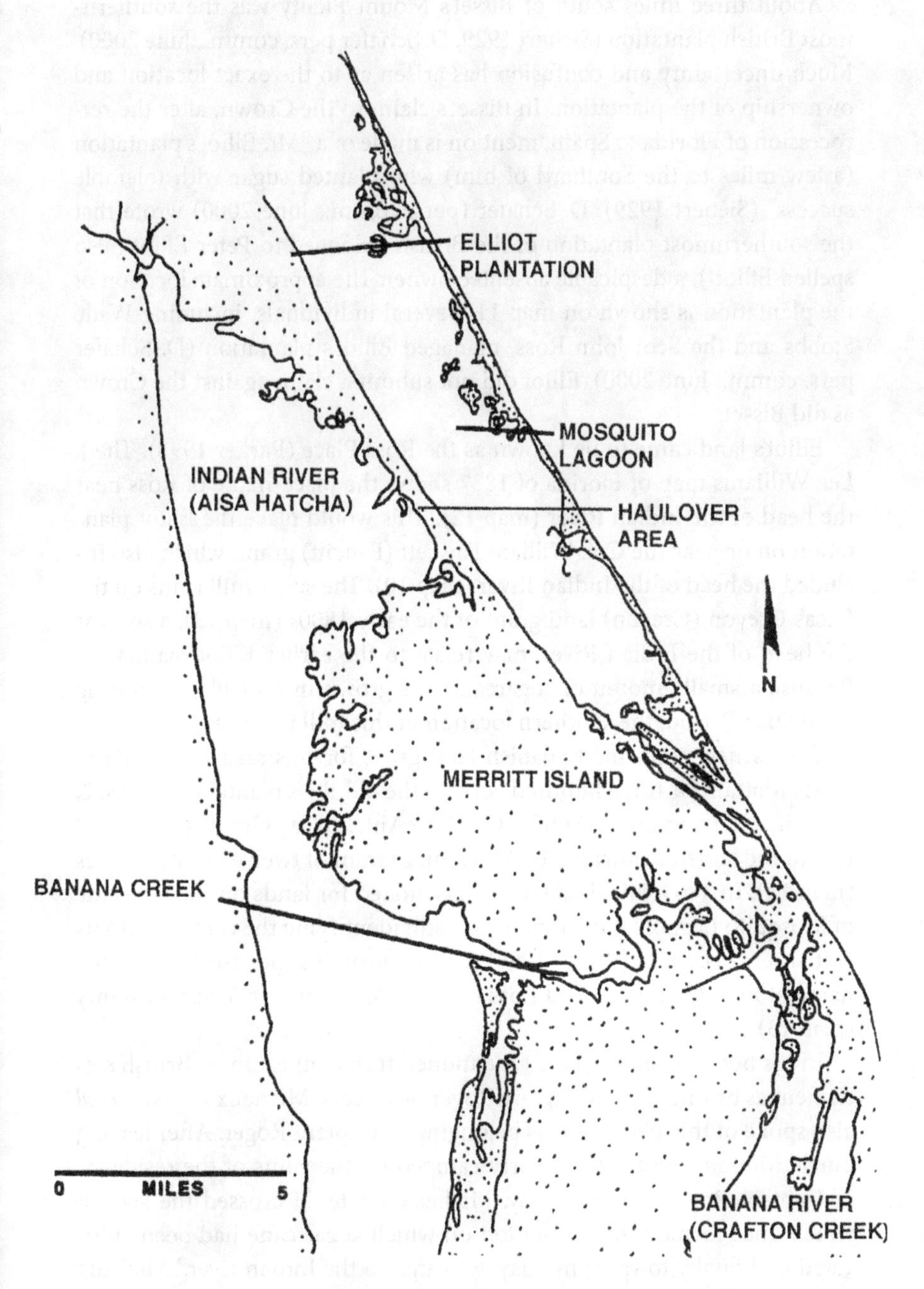

Map 11. Northern section of Merritt Island, Brevard County, Florida.

About three miles south of Bisset's Mount Plenty was the southernmost British plantation (Siebert 1929; D. Schafer pers. comm., June 2000). Much uncertainty and confusion has arisen as to the exact location and ownership of the plantation. In Bisset's claim to the Crown, after the retrocession of Florida to Spain, mention is made of a "Mr. Elliot's plantation (a few miles to the Southard of him) who planted sugar with tolerable success" (Siebert 1929). D. Schafer (pers. comm., June 2000) wrote that the southernmost plantation of the British belonged to Peter Elliot (also spelled Elliott), a despicable absentee owner. The approximate location of the plantation is shown on map 11. Several individuals, including Wade Stobbs and the Scot John Ross, managed Elliot's plantation (D. Schafer pers. comm., June 2000). Elliot did not submit a claim against the Crown as did Bisset.

Elliot's land came to be known as the Ross Place (Parker 1998). The J. Lee Williams map of Florida of 1837 shows the place-name of Ross near the head of the Indian River (map 12). This would place the Elliot plantation on or near the Col. William Fawcett (Faucitt) grant, which also included the head of the Indian River (map 10). The sugar mill ruins on the Lucas Creyon (Crayon) land grant of the early 1800s (map 13), also near the head of the Indian River, may relate to the earlier Elliot plantation. Because a small amount of sugarcane was grown in East Florida during the British Period, the southern location of this mill is significant.

The earliest confirmed Spanish land grant for this area was made to Lewis Mattair for three hundred acres at the old Ross plantation. In 1822, Mattair sold the grant to Antelm Gay. The Antelm Gay Grant is contained within the larger, confirmed H. M. Gomez Grant of twelve hundred acres (map 13). In 1803, Nicolasa Gómez petitioned for lands on the west side of Mosquito Lagoon at the Ross Place, still identifying the acreage with its British-era occupant (Parker 1998). The name of Ross persists today in the name of Ross Hammock and Ross Hammock Mounds in Volusia County (map 13).

In his book, Romans (1775) mentioned this southernmost British settlement as belonging to a Captain Roger or Rogers. Michaux in his *Journal* also spoke of this plantation as belonging to Captain Roger. After leaving Turtle Mound, the Michaux party camped on the ruins of the residence of Captain Roger. The next day, Michaux wrote, "I crossed the swamp which once formed this habitation on which sugar cane had been cultivated and finally towards midday, we came to the Indian river." Michaux remarked that this habitation was the most southern one that the English

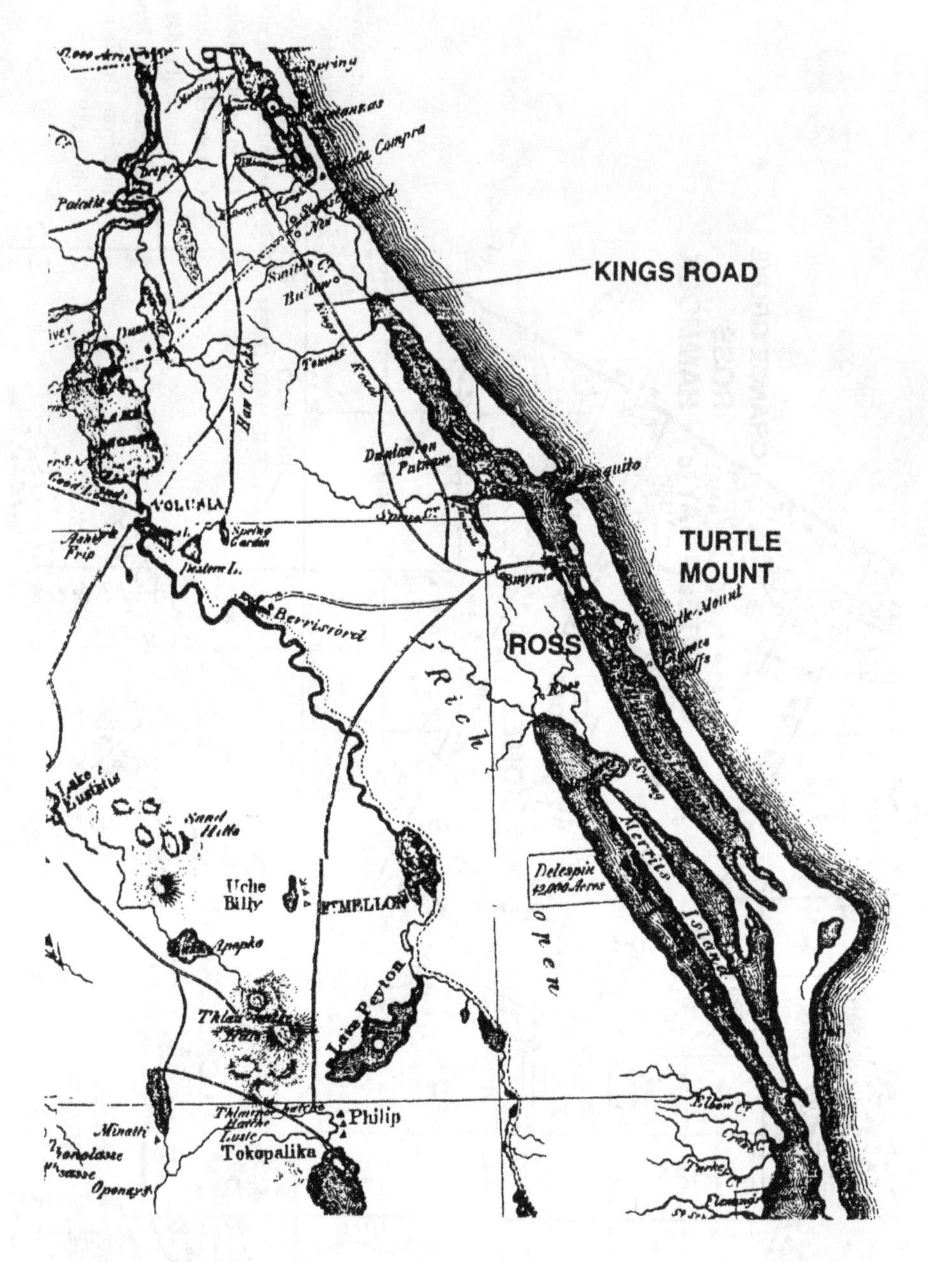

Map 12. Portion of the J. Lee Williams 1837 map of Florida showing the location of the place-name of Ross.

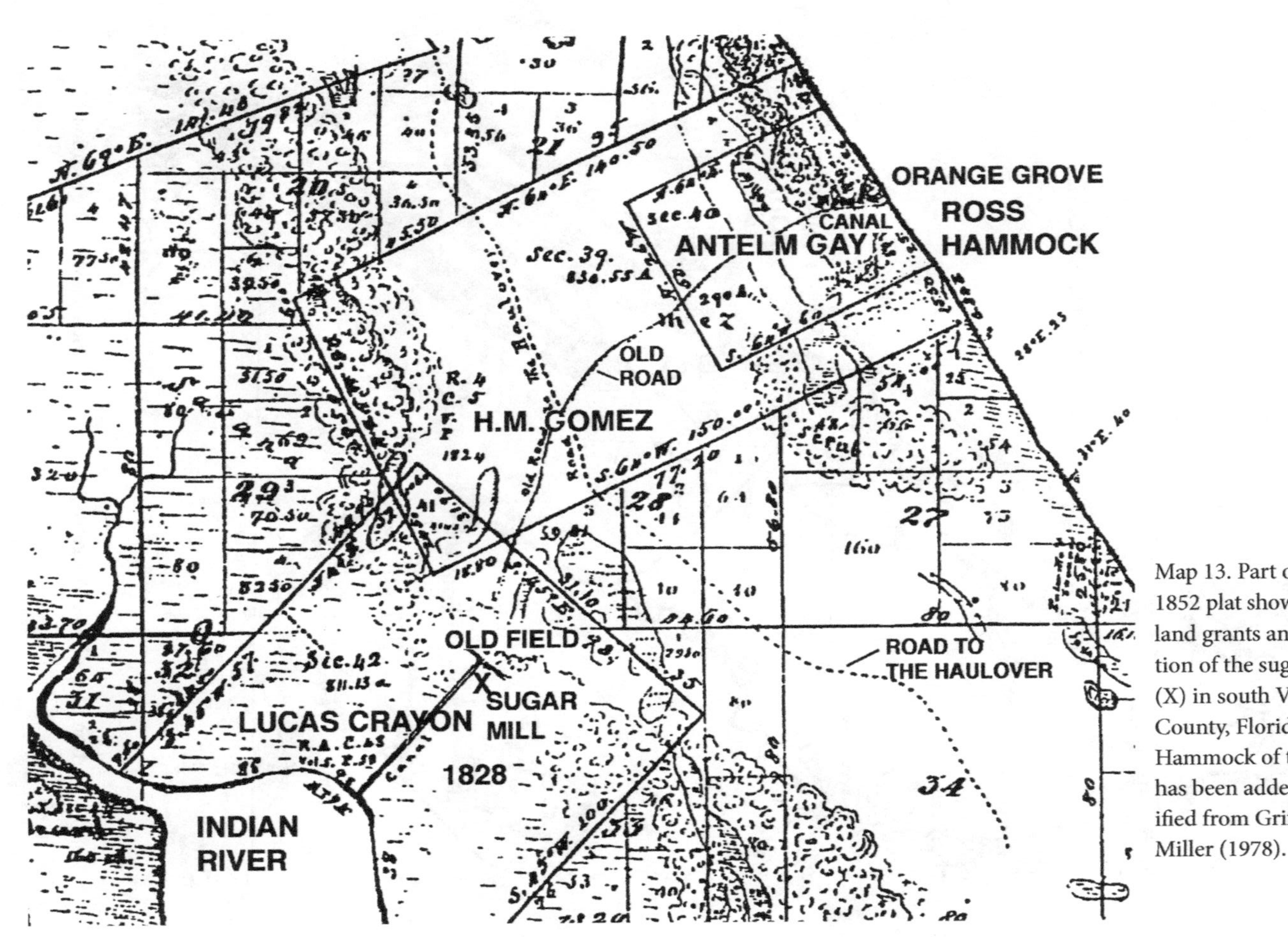

Map 13. Part of an 1852 plat showing land grants and location of the sugar mill (X) in south Volusia County, Florida. Ross Hammock of today has been added. Modified from Griffin and Miller (1978).

had established in Florida. A swamp can be seen on the Lucas Creyon grant (map 13). The Roger confusion may have resulted from a person's poor penmanship or a misprint in Romans's book whereby Ross became "Roger" (Griffin and Miller 1978). Regardless, Mr. Elliot's ownership of the land was contemporary with Captain Robert Bisset's plantation to the north (Watson 1927; Griffin and Miller 1978).

Two towns, New Smyrna (maps 7–9) on the Hillsborough River (now North Indian River) and west of Mosquito Inlet (now Ponce de León Inlet) and the older Rollestown (maps 1 and 3), also called Charlotta (also spelled Charlottia and Charlotia) after Queen Charlotte, on the St. Johns River, emerged during the British Period (maps 1 and 6). Both enterprises collapsed; present-day New Smyrna in Volusia County has its roots in the old settlement founded by Dr. Andrew Turnbull.

Turnbull, a wealthy, Scottish physician who had lived and traveled in England and the Mediterranean area, had a vision of establishing a British colony in East Florida. Turnbull's settlement, called New Smyrna, was the most ambitious operation attempted in Florida at that time (Corse 1967). Initially New Smyrna was the largest British colony established in the New World, greatly exceeding Jamestown (Panagopoulos 1966; Corse 1967). Governor Grant reported that Turnbull's operation was the largest importation of white inhabitants ever brought into America at one time. Turnbull's vision was so convincing and attractive that it gained support of prominent Londoners of financial and political means as well as Governor Grant, who took an immediate fancy to the Scot's idea. The Turnbull colony would be composed of indentured servants transported from the Mediterranean area to Florida, where the maritime climate was thought to be similar.

Dr. Turnbull and Sir William Duncan each received twenty thousand adjoining acres in land grants (Corse 1967). Then Sir Richard Temple's twenty thousand acres were added to the New Smyrna colonial holdings. In addition to these tracts, five thousand acres of land were obtained for each of Turnbull's four children (Siebert 1931; Mowat 1943). Turnbull's settlement ultimately reached 101,400 acres (Siebert 1929; Panagopoulos 1966; Corse 1967).

Having an order in council for twenty thousand acres of land in East Florida, in June 1766, Turnbull with his family sailed from England and arrived at St. Augustine in November (Siebert 1929). After consulting with the government surveyor as to where the best lands in Florida were located, he chose the acreage for his settlement to be in the Mosquito District, west of Mosquito Inlet. Dr. Turnbull named the settlement New Smyrna in honor of his Greek wife's birthplace of Smyrna, Greece (Strickland 1965; Panagopoulos 1966; Corse 1967). The Turnbull acreage had some of the most valuable land, both in beauty and fertility, that existed in the British province. Turnbull's own residence, about four miles from the New Smyrna settlement, was on a personal grant of three hundred acres that adjoined the larger grants (Siebert 1929; Griffin 1991).

Turnbull returned to England in late March 1767. Later in the year he sailed to Greece, Leghorn (Livorno, Italy), and Minorca to acquire settlers for the forthcoming colony. On June 26, 1768, Turnbull arrived in St. Augustine with the first four ships from the Mediterranean area bound for the newly proposed settlement. In July the other four ships arrived (Panagopoulos 1966; Rasico 1987). A total of 1,403 individuals, three hundred families, was lured to the New World (Hanna and Hanna 1950; Mowat 1943; Panagopoulos 1966), but only 1,255 individuals survived the rough trip across the Atlantic (Rasico 1987). Greeks, Corsicans, Italians, French, and Spanish came, but the largest number of settlers, probably no less than one thousand, were Minorcans from the then British-owned Mediterranean island of Minorca. A severe famine that existed in Minorca at the time provided the impetus for many to leave their stricken homeland (Griffin 1991).

Members of the New Smyrna colony cleared land, built drainage canals, and produced crops of indigo and corn. Mulberry trees were planted for silkworm culture and silk production, grapes for wine, and nonnative cochineal insects were imported for their red dye (Corse 1967; Brewer 1988). Only indigo and corn were successful. Though planned, rice and silk were never produced for commercial purposes (Panagopoulos 1966). The overwhelming success of indigo production, a very labor-intensive operation, was probably the reason little effort was spent on other crops (Griffin 1991). Additional income for the colony came from production of tar, pitch, turpentine, and barilla.

Andrew Turnbull was proud of his New Smyrna; however, his enterprise collapsed. The majority of the colonists revolted and abandoned the settlement in the summer of 1777; six hundred individuals went to St. Augustine (Griffin 1991). Governor Tonyn took it upon himself, without consulting Turnbull, who was away, to cancel the indentures of the colonists.

The collapse of the New Smyrna colony is often attributed to cruel and barbaric treatment of the colonists by Turnbull and his overseers. Bernard

Romans (1775), a hostile critic, was one of the first writers, if not the first, to expound on the barbarism of Turnbull, but his account evidently is exaggerated and biased (Corse 1967). Though harsh and inhuman treatments took place in which many colonists were beaten, overworked, and underfed, other factors contributed to the colony's demise (Panagopoulos 1966; Corse 1967; Rasico 1987). From the very beginning, there had been inadequate drinking water, clothing, living quarters, and meat and fish in the diets; the number of colonists was double that originally anticipated. Diseases such as malaria and scurvy prevailed; the former probably accounted for more deaths than any other disease. The hot and humid climate of Florida was harsh compared to that of the colonists' homelands. The British government offered little protection to the colony from Indian raids and other threats. Lastly, serious political and personal feuds drove a wedge between Turnbull and his associates and Acting Governor John Moultrie, and later Governor Tonyn.

Tonyn certainly accelerated the demise of the colony in several ways (Panagopoulos 1966). For instance, he spread a rumor among the colonists that Turnbull was going to take the land that he had promised them (Corse 1967). Captain Bisset alleged that Tonyn dismantled the colony to get recruits for his corps of Rangers (Siebert 1929). In contrast to Moultrie and Tonyn, Governor Grant had supported Turnbull's endeavors. Grant, Chief Justice Drayton, and others, including Moultrie himself in 1773, stated that Turnbull was devoted to his colony and that he did his best (Corse 1967). James Penman, Chief Justice Drayton, and Captain Bisset all agreed that Tonyn broke up the colony (Corse 1967).

After nearly two years under custody of the provost-marshal, Turnbull left Florida upon his release from jail. He, his wife Maria Gracia Dura Bin (ca. 1730–1798), and James Penman took a ship to Charleston, arriving on May 13, 1781 (Panagopoulos 1966). James Penman left Charleston and later moved to London, where he became a merchant. In May 1786, he became Turnbull's lawyer and presented claims pertaining to his losses in East Florida (Siebert 1929; Panagopoulos 1966; Corse 1967). Turnbull's friend William Drayton became a district judge of the United States in Charleston, having returned to his family estate, Magnolia Gardens, then known as the Drayton House, on the Ashley River outside of Charleston. Turnbull practiced medicine and became one of the earliest members of the South Carolina Medical Society. At age 73, Dr. Andrew Turnbull died on March 13, 1792, and was buried in Charleston's St. Philip's churchyard cemetery in an unmarked grave. His wife died at age 68 on August 2, 1798;

a stone marks her grave at St. Philips (Siebert 1929; Panagopoulos 1966; Corse 1967).

The Minorcan contributions to the British province and, later, to the Spanish were significant. Many became farmers, fishermen, hunters, and craftsmen. Soon after his arrival Spanish Governor Zéspedes wrote that the Minorcans were an industrious people (Griffin 1991). Descendants of the New Smyrna colony still reside in St. Augustine and other places in Florida.

André Michaux camped on the ruins of Turnbull's property in New Smyrna on Easter Sunday, March 23, 1788. The account in his *Journal* of Turnbull's deplorable and barbaric behavior toward his colonists was probably based on his reading of Romans's (1775) book. Michaux stayed for five days at the house of an unnamed Minorcan located on the Northwest River (Pellicer Creek of today). This Minorcan loaned Michaux three horses.

During the British occupancy of Florida and beginning with the governorship of Grant, roads were built or expanded, including the famous Kings Road (map 12). This road, built in sections, was the first major land route along Florida's east coast. The road from St. Augustine to Turnbull's property at Mosquito Inlet was blazed in December 1767 (Adams et al. 1997). In late 1774 this southern segment of the road, linking St. Augustine with New Smyrna, was completed. Captain Robert Bisset was in charge of building an extension from New Smyrna south to Mr. Elliot's place (Parker 1998). Michaux probably used the road at least from today's Pellicer Creek to the Tomoka River.

The population of East British Florida increased, and trading with the Indians was well established by James Spalding and prosperous merchant Donald McKay (also spelled Mackay). With McKay's death in 1768, Roger Kelsall became Spalding's partner (Coker and Watson 1986).

The two most famous Indian trading posts of East Florida were located on the St. Johns River: Spalding's Upper Store near Volusia in Volusia County and Spalding's Lower Store at present-day Stokes Landing, about six miles southwest of Palatka in Putnam County (map 3). Some disagreement exists as to the exact location of Spalding's Upper Store. Mrs. Lillian Dillard Gibson, curator of the Volusia Museum at Volusia, Florida, has strong documentation, including a description by Fernando de la Maza Arredendo (dated March 1, 1817, St. Augustine) and a survey map (dated June 8, 1817) by Andres Burgevin, surveyor appointed by the government, that shows the store situated on one of several large shell mounds on the east bank of the St. Johns River at Volusia. Because of this evidence, Coker and Watson (1986) place the store on the east side in their published map. Slaughter (1996a, p. 103) also shows the Upper Store on the east side of the river in his map. Cartographer William Gerard De Brahm, on his 1769 map (Kings MSS 211.12, British Library), clearly shows the store on the west side of the St. Johns River at today's Astor, Lake County, and so does Francis Harper (1958), editor of Bartram's *Travels*. The De Brahm map shows all of the Spalding's trading stores on the St. Johns River to be located on the west bank. We have chosen to place the Upper Store on the western bank of the river (map 3).

William Panton, a dedicated Tory and senior partner of the firm, and Thomas Forbes organized the firm of Panton, Forbes and Company, an Indian trading business. By the end of the American Revolution the firm had become the much larger Panton, Leslie and Company, the best known of the Indian traders in the Southeast (Coker and Watson 1986; Mowat 1943), and by 1787 it had acquired from Spalding the Upper and Lower Stores and other Indian trading posts. Both William Panton and John Leslie were born in Scotland, but moved to Charleston. From there they came to Florida in 1765. William Panton settled at Pensacola, but John Leslie remained in St. Augustine to supervise the business of the East Florida stores. After Florida returned to Spanish control, Panton, Leslie and Company retained the monopoly on trading with the Indians (Mowat 1943).

André Michaux visited John Leslie on April 18, 1788, to discuss traveling in Indian territory. The next day Leslie invited Michaux to dinner at his house, located along the waterfront in St. Augustine. This dinner engagement occurred before Michaux had made his trip on the St. Johns River. During that later exploration, the Upper Store and perhaps the Lower Store were in operation (Tanner 1963).

It was during the British occupation of Florida that John Bartram, age 66, recently appointed King's Botanist to George III, and his son, William, age 26, came to Florida. This Florida trip was the last major expedition taken by the elder Bartram, who was in poor health and going blind (Porter 1993). The Bartrams stayed in St. Augustine from October 11 to December 19, 1765, visiting Governor Grant and botanizing in the area (Berkeley and Berkeley 1982). On November 15–18, 1765, the Bartrams attended the first congress at Fort Picolata. The congress, attended by chiefs of the

Creek Indians, Colonel John Stuart (d. 1779) of Georgia, superintendent of Indian affairs, Governor Grant, and the Bartrams, was held to fix the boundary between the Indians and British in East Florida.

Fort Picolata was located on the east bank of the St. Johns River, about twenty miles west of St. Augustine and about twenty-five miles north of Rollestown (map 2). The fort originally was a mission station (Sastre 1995). Beginning in 1670, when the English sought to control Florida, the Spanish constructed defensive posts along the St. Johns to safeguard the river's crossing, the Spanish missions, and St. Augustine's western flank. Around 1700 a wooden fortification at Picolata was built by the Spanish. Lieutenant George Dunbar, under orders of Georgia's governor, General James Oglethorpe, attacked the fort on December 28, 1739. Picolata was reduced to ruins by mortar shells and later burned by Dunbar's Indians. The fort, rebuilt in 1755, had a three-story square tower made of coquina rock, a natural shell deposit taken from Anastasia Island up the St. Johns River by barge (Mowat 1943; Sastre 1995). The blockhouse was enclosed by a square wooden palisade surrounded by a square moat on the north, east, and south sides (Sastre 1995). Following the two congresses held at Fort Picolata, the structure was abandoned in 1769. It was not until Governor Zéspedes's successor, Governor Juan Nepomuceno de Quesada, that repairs of Fort Picolata were begun. Today, no trace of the structure exists and the exact site where the fort stood is not known (Mowat 1943; Sastre 1995).

After the first congress terminated, the Bartrams returned to St. Augustine, staying until December 19, when they left the city to travel the St. Johns River, mapping it, looking for its source, discovering new plants, and extracting information that would help the British Empire exploit their American possession for commercial development (Slaughter 1996a; Hetrick 2000). The St. Johns River was to be the central highway for a new agricultural enterprise for the British.

On February 13, 1766, the explorers were back in St. Augustine. John sailed for Charleston in March; however, William, much to the displeasure of his father, stayed in Florida. He acquired, with the support of his very reluctant father, a five-hundred-acre land grant. The plantation was at Little Florence Cove, St. Johns County, on the east bank of the St. Johns River, an exact location discovered only recently (Schafer 1995). There, William would become a Florida planter, assisted by six slaves bought by his father, growing indigo and rice. This enterprise was abandoned in the fall, lasting less than a year; William Bartram had once again become a failure in the eyes of his father. The young man simply had no skill for or interest in business. After his unsuccessful attempt at being a planter, William worked for cartographer William Gerard De Brahm, surveying Andrew Turnbull's model plantation of one hundred thousand acres at New Smyrna along the Mosquito River. It was during this time in November 1766 that William Bartram was shipwrecked off the coast near today's Ponce de León Inlet (Harper 1958; Derr 1998). Bartram returned to Philadelphia in the fall of 1767.

Then, on March 20, 1773, with financial and psychological support of Dr. John Fothergill (1712–1780), a noted English Quaker physician and owner of one of England's largest private botanical gardens, thirty-five-year-old William left Philadelphia to begin his famous Southeast explorations that would last nearly four years. With this venture and with the publication in 1791 of his *Travels through North and South Carolina, Georgia, East and West Florida*, William Bartram finally achieved success—a success far greater than he would have gained as a printer, painter, or planter.

Spanish Florida: 1784–1821

Charles III of Spain declared war against Britain on June 21, 1779. Baton Rouge was captured during that year and soon thereafter Mobile fell to the Spanish. With seizure of Pensacola in May 1781, the largest battle ever fought in Florida, the demise of the British Floridas had arrived. Even East Florida, where little fighting occurred, was also returned to the Spanish. The British government considered both Floridas disposable and their return to Spain of little consequence (Fabel 1996). This change of ownership affected the lives of thousands of Loyalists and nonloyalists alike who had recently fled from the United States to East Florida to begin a new life. The control of Florida by Great Britain, which had lasted almost twenty years, came to an end when Spain, Great Britain, and France signed the Second Treaty of Paris in 1783.

Life in East Florida was suddenly changed with the Second Spanish Period (1784–1821); however, English interest and influence in East Florida were never completely removed until the United States took possession of Florida in 1821 (Tebeau 1980). The division of Florida into East and West colonies, created by the British, was retained by the returning Spanish.

Spain now occupied more territory in North America than in previous times; however, Spain was weaker than it had been twenty years earlier when Florida was lost to the British (Tebeau 1980).

Vizente Manuel de Zéspedes y Velasco (1720–1794), age sixty-four, acknowledged his appointment as governor and captain general of Spanish East Florida while in Havana. A native of Castile, Spain, accomplished military leader, and interim governor of Santiago, Cuba, Zéspedes left Havana in June 1784 with fifteen vessels, five hundred troops, and members of his staff and family (Tanner 1963). The Zéspedes party arrived in St. Augustine on June 27, 1784. His wife, Doña Maria Concepción Aróstegui, of Basque origin, and their two daughters, Dominga and Josepha, came with the newly appointed governor. Two sons, Vizente Domingo and Antonio, were in the Havana Regiment, soon destined for St. Augustine. Two older sons, Fernando and Thomas, were with army regiments in Spain (Tanner 1963). For the next six years St. Augustine would be home to Vizente and Doña Maria. His governorship terminated in July 1790, when they returned to Havana, where Zéspedes died four years later. Zéspedes was buried in the convent of San Francisco in Havana (Tanner 1963).

Prior to the Spanish takeover of East Florida, the population of six thousand persons had increased to sixteen to seventeen thousand or more individuals. Both Loyalists and nonloyalists had moved from Savannah, Charleston, and other areas, anticipating that Britain might retain the colony. But when Zéspedes arrived at St. Augustine most of the British residents, including old folks, refugees, whites, and Negroes, had already left Florida, bound for neighboring states, England, Nova Scotia, the West Indies, Bermuda, or the British-owned Bahamas (Troxler 1989; Gannon 1993). Contrary to earlier written accounts, however, a total evacuation of British residents did not occur with the change of guard. In fact, more than five hundred East Florida Loyalists remained and served under the Spanish government (Troxler 1989). In spite of the British presence, East Spanish Florida, excluding St. Augustine, was solidly a Spanish-oriented bloc with only 11 percent of the population being Anglo-Protestant (Johnson 1989).

It was not until November 1785 that ex-Governor Patrick Tonyn and the last British fleet were headed back to England. During Zéspedes's first sixteen months as governor, Tonyn was a constant thorn to the Spanish government and to Zéspedes.

Within a short period, the East Florida population in 1786 had dropped to about three thousand individuals. St. Augustine's population numbered

772, exclusive of the military personnel (Johnson 1989). Although these figures represent a decline from the British time, St. Augustine was never more diverse. The polyglot population consisted of Minorcans, Italians, Greeks, French, British, Swiss, Canary Islanders, Indians, Negroes, Cubans, Americans, and Spaniards, including civilian and military personnel. Compared to the city, the rural population was more homogenous, overwhelmingly Protestant, and mostly farmers (Parker 1990; Coker and Parker 1996).

The dominant society of St. Augustine consisted of the Spanish administration, Cuban civil servants, Floridano families, and Minorcans. The Minorcans, who had come earlier to St. Augustine from Turnbull's New Smyrna colony, comprised 469 persons, most of the town's population (Johnson 1989). These colonists were glad to see the British domination gone and the Spanish in charge. Former Spanish residents who had left during the British takeover returned to reclaim their land; they found little change to the town, but the colony was different from the one they had left twenty years earlier. These so-called Floridanos totaled about 216 persons (Johnson 1989). Outlaws and their gangs, like the infamous James and Daniel McGirth (also spelled McGirt and McGirtt), continued to roam the countryside, stealing cattle and slaves and causing unrest as they had done during the British times. Daniel McGirth had been a scout for the Americans at the beginning of the Revolution. A Whig officer stole his horse; while McGirth was in prison, the British whipped and badly mistreated him. Because of this the McGirth brothers detested the British. They desired to gain control of north Florida for the Americans. Of course, land-seeking Americans also moved into Florida (Tebeau 1980).

Former British Loyalists outnumbered Americans, both as inhabitants of and immigrants to St. Augustine (Johnson 1989). Zéspedes made it clear that Americans or residents from the north were not at all welcome in East Florida. However, the inhabitants along the northern riverbanks of East Florida were of American birth. The governor was interested in population growth, re-establishment of the Roman Catholic Church—which had greatly weakened since the earlier days of Spanish rule—land development by the Spanish, using land-grant procedures similar to those established by the British, and, in short, the development once again of a Spanish population in St. Augustine and East Florida. Still, the Spanish government allowed most British who remained to retain their lands and slaves provided they met the royal requirements, including taking an oath of loyalty to Spain (Johnson 1989; Sastre 1995). Not all who stayed converted to Catholicism. Many wealthy British landholders remained in East and West Florida after Florida reverted to Spanish control. One such person who continued to operate his land under Spanish rule was Francisco Philipe Fatio, a Swiss from Berne, who had a large, premier plantation called New Switzerland located on the eastern bank of the St. Johns River, north of Fort Picolata (map 2). Fatio had arrived in British East Florida in 1771, and had received the ten-thousand-acre plot during the administration of Governor James Grant. New Switzerland was destroyed by Indians in 1812 (Parker 1990).

Many of the plantations, especially those located along the east coast, were vacated because the Spanish initially took no interest in continuing these British landmarks. It was not long before these formerly occupied lands were overgrown with a mixture of native and naturalized flora. The southern extension of the Kings Road built by Captain Robert Bisset during Governor Grant's administration also became overgrown because of neglect (Mowat 1943). Dr. Turnbull's New Smyrna and Denys Rolle's Rollestown were no longer in existence. Many of the mansions, outlying buildings, and other facilities were destroyed by Indians. Perhaps the destruction of Bella Vista and other plantations slowed the desire of the Spanish to continue immediately the same regime of cultivation (Forbes 1821). It was not until 1787 that the Spanish decided to repopulate the colony. As a result, a vigorous, renewed interest in cultivation and plantation development emerged in Spanish Florida. Grants of land were offered to immigrants, including non-Spaniards.

His military background and experience and an honorable career in the royal service to Spain contributed to Zéspedes's appointment as governor. A man of mild temperament, Zéspedes was concerned with the well-being of the people living under his jurisdiction, but throughout his governorship, he lacked funds for administrative affairs and for the promotion of public affairs and private industry (Lockey 1949). Zéspedes constantly had to beg Havana for the funds he did manage to obtain.

Zéspedes had a competent, though small, administration. His immediate personnel included Irish-born Captain Carlos Howard, secretary and administrative assistant; Gonzalo Zamorano, native of Castilla la Vieja and director of the royal treasury; and Mariano de la Rocque (also spelled Roque), chief engineer. La Rocque made a detailed map, dated April 25, 1788, of St. Augustine, showing residences, government and commercial facilities, churches, and other landmarks (map 14). He also supervised construction of the Spanish cathedral. Fathers Miguel O'Reilly and Thomas Hassett, both Irish, served the religious affairs and needs of the Roman Catholics. Father Pedro Bartolomé Camps, whose residence extended back to the British Period, was a native of Minorca and priest to the Minorcans at the New Smyrna colony and at St. Augustine (Panagopoulos 1966). John Leslie, of the Panton, Leslie and Company that was involved with Indian trading under the British, still lived near the waterfront in St. Augustine. The company continued to operate during the Spanish rule.

Perhaps the most brilliant, discreet, trustworthy, and capable person of Zéspedes's administrative staff was Captain Carlos Howard. Howard was Zéspedes's main assistant, serving as secretary of state for eleven years at St. Augustine (Lockey 1949; Dovell 1952). Not only did Howard possess military prowess, but he also knew English and French as well as Spanish (Lockey 1949; Tanner 1963). Because Zéspedes did not know English, he relied heavily on Howard's skills. Carlos Howard was one of the first individuals whom Michaux met when he arrived in St. Augustine. Howard served as secretary of the government, translator, and member of the junta de hacienda of Governor Juan Nepomuceno de Quesada, who succeeded Zéspedes. It was Howard who was placed in charge of the Georgia-Florida incursions in 1794 when the French planned an attack on both East and West Florida (Tanner 1963). In 1796 Howard was promoted to lieutenant colonel of the Louisiana Regiment and became military commander of Upper Louisiana (Tanner 1963).

4

André Michaux Comes to Florida

1788

This translation of André Michaux's *Journal* recounts his stay and explorations in East Spanish Florida. Notes providing clarification and amplification of the plants, people, and places mentioned in the *Journal* have been added at the end. The superscript numbers in the *Journal* correspond to the numbers in the notes.

Sargent's (1889) transcription of Michaux's *Journal* has been used primarily. His work was based on Michaux's original notebooks, located at the American Philosophical Society (APS) in Philadelphia. In addition to Sargent's work, the copies of William Spawn, also located at APS, and that of Marie-Florence Lamaute obtained from her were consulted and compared with the Sargent account. The original punctuation, capitalization, and plant names for the most part have been retained. Scientific names are not in italics. The original place-names are retained; however, most are spelled as they are today.

3rd Notebook (Cahier). 1788

JOURNAL

Thursday 14 February 1788. I prepared myself to embark. I bought a small negro male for the price of 50 pounds and I rented another one for one shilling a day. Embarked at 12:30 p.m. for St. Augustine in Florida.

The 15 calm weather and contrary winds; we stayed anchored inside the Charleston Bar.

The 16 a strong wind came up during the night, several ships were dragging their anchors. A schooner came and hit the one on which we were, but without any damage. We managed to separate them. Then came the rain, and we hoped that the wind would turn from the south to the north, but it continued and that evening we obtained shelter from the wind below Sullivan Island, in view of Charleston.

Sunday 17 February 1788 we stayed at anchor and I went to look for plants on the island of Sullivan. I recognized only a few plants which were worthy of being noticed, because this little island, besides being sterile enough due to its exposure to the winds is burned every year according to the custom of the Americans who annually set fire to all the forests. The English, during the last war cut all of the big Chamaerops; all that's left are young ones that do not have fruits; I also noticed a shrub whose fruits indicates it to be a Croton and a grass (gramen).

The 18 the wind calmed down but it still was not favorable.

The 19 we raised anchor and we went beyond the Bar, but the contrary winds forced us to return [to the Bar].

The 20 we sent a canoe to the city and I took advantage of the occasion. I went there not only to renew our provisions consumed during the bad times spent due to the contrary winds, but also with the hope of hearing some news from France by way of New-York which should have already arrived. There was a schooner destined for New-York and I regretted all the more bitterly the eight days lost due to the contrary winds; if I had stayed in Carolina I would have been able to prepare a shipment for the 24th of February which was the day the schooner was to depart and likewise if the wind had been favorable to go to St. Augustine I would have been able to send a very interesting shipment by the schooner, whose departure date had been established for the 24th of February from Charleston and which could have arrived easily before the 10th of next March in time for the departure of a steamer headed for France.

The 21 we still remained anchored and in the evening there arose a heavy wind accompanied by rain.

The 22 the sea's turmoil and the wind having stopped, we were hopeful of having the North Wind which would have been favorable for us.

The 23 the wind was very favorable, but we spent all day trying to pull up our anchor which had become so stuck that we were about to abandon it, but toward the evening, with the help of another ship which was stronger, we were able to pull it up.

Sunday 24 February 1788, we set to sail with a very feeble wind, but favorable enough.

The 25, we encountered a southly wind which was contrary; it lasted until the next morning.

The 26 & the 27. we stayed at sea and finally toward the evening, we observed the Florida coast.

The 28, we entered the Port of St. Augustine and we disembarked at 1 o'clock in the afternoon.[1]

There came on board some officers of the Government who asked what I came to do and if I had brought any merchandise: I responded by saying that I came only to observe the natural history of Florida and that I had already obtained permission from his Excellency the Governor. Immediately they told me that it was necessary to present myself to him. I told the governor that I had no other object than natural history and when I would be prepared to visit the different parts of the country, I would inform his Excellency and I would give him acknowledgment of the most interesting Discoveries.[2]

He told me that I was welcome and that all of the services he could render, he would give me. He showed me every courtesy and had a message sent to the place where I had taken lodging so that I might be graciously attended.[3]

The 29, the day was spent in visits.[4]

The 1st March 1788, I went to look for plants and I recognized an Andromeda of a new species No. 1, 2, & 3.[5]

Sunday 2, we went to Church and we heard Mass at which his Excellency the Governor was present.[6]

The 3 Thermometer at 9 Degrees Réaumur[7] above zero at 6 o'clock in the morning. We went a distance of 5 miles, but a storm accompanied by thunder and lightning, went through us and soaked us completely, and we came back without picking up any interesting plants.[8]

The 4, the wind accompanied by rain lasted all night, the thermometer was at 5 1/2 degrees. The storm was little less violent during the day; we went a distance of more than 6 miles and saw only the interesting shrubs found the 1st of March, namely No. 2 &

No. 3. I also collected an unknown shrub that looked like an Andromeda No. 4 but which differed totally in its fruits.[9]

The 5 Wind from the North West, Thermometer 2 1/2 above zero in the morning. The day was used to read the description of Florida and to verify this description with a map which had been lent to me.[10]

The 6 consulted several inhabitants on the means of going to look for plants in the south of the Province.

The 7 I bought a small canoe and hired two men who could maneuver it.[11]

The 8 I bought some provisions for the voyage and a lot of powder and lead to be able to kill some game, because the areas where I was going to visit were uninhabited, and only frequented by Indians.

The 9 Sunday settled all things for the voyage. The thermometer in the morning was at 5 degrees.

The 10. Thermometer in the morning was at 5 1/2 Degrees. Wind from the North West. A worker was occupied sewing the sail of the canoe and making some repairs.

The 11. Thermometer in the morning at 4 3/4 Degrees above zero. Wind from the North West. The sail and other parts of the canoe not having been ready, I went to visit the property of a local person so that I could establish there a depot for trees.

Wednesday 12 we left St. Augustine in our canoe which had five persons namely my son & me, two oarsmen and the negro whom I had brought from Charleston. The wind was favorable, but the tide that was against us formed waves which came into our canoe and we decided to stop at the house of a respectable old man established here for 52 years on the isle of St. Anastasia. This man being the most hard working and the most industrious man in all of Florida had turned his place into a Paradise notwithstanding the different pillages by Pirates to which he had been exposed and the revolutions which he had twice experienced by the changing of ownership, this Province having gone since his day from the Spanish, to the English, and once again to the jurisdiction of the Spanish.[12]

Thermometer in the morning at 12 Degrees.

The 13 we went around the isle of St. Anastasia; we stopped at around 14 miles from St. Augustine and I recognized on the shore two species of trees * * * called by the English Mangrove and in some parts of this island, the Zamia.[13]

We arrived in the evening at fort Matanzas situated on this island. I used the few hours remaining to me to look for plants within a short distance from this fort.[14]

The 14 we tried to go past la barre de Matança 20 Miles from St. Augustine, where the island of St. Anastasia ends, but the wind which came from the sea formed waves that filled our canoe; we decided to stop at the house of a Minorcan who lived 3 miles distant at the mouth of the North West river and 24 miles from St. Augustine.[15]

Thermometer 14 Degrees.

The 15 the wind still coming from the sea kept us at the house of the Minorcan. I visited the surroundings and I recognized only the same plants known to me from Carolina and from Georgia, namely: Magnolia grandiflora, Quercus phellos, Pinus taeda, Myrica cerifera, Bign. sempervirens, Juglans hickory.[16]

Sunday 16 March, thermometer in the morning at 14 Degrees. We have taken a horse and a guide to go up the river of the North called the North West river. We traveled 22 miles and noticed only, other than plants common to Carolina and Georgia, such as the Magnolia grandiflora, Gordonia lasianthus, Acer rubrum, Laurus borbonia, Cup. distic., Myrica cerifera etc etc. In addition to these trees I also saw along the river, which should not be called a river but a stream, Andromeda arborea, Zamia pumila, Chamaerops repens and a leguminous shrub with ternate leaves No. 17 and another shrub unknown No. 18, a Halesia tetraptera with small flowers and two species of Annona & etc.[17]

The 17. we continued to follow this river, within a short distance I saw Viburnum cassinoides, Ziziphus scandens, Lupinus pilosus flore ceruleo.

I gathered a lot of seeds from shrub No. 17 and a new Andromeda. Finally seeing that the soil was always arid without interesting plants I decided to turn back.[18]

The 18 I collected no new plants, but I recognized on the bank of the North West river and along the Matanzas River an Andromeda with almond-shaped leaves around 10 to 12 ft in height, it formed hollow stalks which were very straight from which the Indians, so it is said, use for their Pipes. I didn't see it flowering, but I believe it is the one that Bartram designated to me under the name of Andromeda formosissima.[19]

The 19 the two oarsmen whom I had sent with my negro, not having given the signal upon which we had agreed, I decided to go to that place and there I learned from a soldier of Fort Matanzas that they had found the wind to be favorable to cross la Barre and that the tide had forced them to leave without having the time to light the fire, the signal we had agreed to give. Coming back I went by a place abundant with oranges and two miles further I found several interesting shrubs.[20]

The 20, the Minorcan at whose house we had been staying, gave me three horses so I could go join the Oarsmen, because the Sea was so rough at la Barre de Matança that it would have been imprudent for us to pass with our Baggage.[21]

We left at 7 o'clock and we walked until 6 o'clock in the evening without stopping. I saw the most arid countryside in Florida, during this walk, with the exception of one Plantation at which we arrived at 5 o'clock in the evening, that had belonged to Governor Moultrie at the time when the English possessed Florida. Finally at 6 o'clock we arrived at the mouth of Tomoka Creek and we camped on the Bank of the Lagoon. (This is a canal formed by Islands which are strung out along the coast of America.) Where these Islands are interrupted then the Sea comes crashing against the river's edge and sailing by Boats is dangerous when the Wind comes from the Sea. One can navigate with small Boats from the Carolinas to Cape Florida and this Navigation is called inland navigation (Navigation de l'Interieur) and the different arms of the Sea formed by these Islands, which extend [along the coast], are named Lagoons which take their names according to the areas and the islands that they enclose. We fired a gunshot and our oarsmen responded quickly by firing another gunshot. They had arrived the evening before without any danger except that their canoe overturned twice by the waves and thus they got wet, but they were very experienced.[22]

We were then about 40 miles away from St. Augustine in a straight line and a mile from the mouth of the Tomoka Creek.[23]

The 21 we passed over to the left bank of this Lagoon where there was an abandoned habitation. I saw Orange trees covered with fruit and I picked up several interesting shrubs. We came to camp that evening on the Island of the Orange Trees 4 miles away from the habitation of Mr. Penman which was abandoned. In the interval, we visited also several houses which had been abandoned and which were numerous enough to have been called a Village.[24]

The 22 we experienced a considerable rain which had started during the night and which lasted until midday.

Our navigation was about 6 miles and we camped on dry ground at 4 Miles distant from the mouth of Spruce creek. There I found Carica papaya.[25]

Sunday of Easter, 23, the wind was favorable enough and we came to camp between la barre de New Smyrna and the ruins of this town which had been founded there in the time of the English. This establishment had been built by Doctor Turnbull at the expense of the Company of which he was the director. More than 1200 people, men women and children, the greater part of them from Minorca, had been enticed from their country. The harshness and the oriental Despotism with which this barbarian led his Colony was still the subject of conversation of the inhabitants of St. Augustine when I was there. This place is designated in a New Map of Florida published in London several years ago by the name of côte des mosquitos.[26]

The 24 thermometer Rheaum. at 7 Degrees above zero, Wind from the North West very noticeable.

We came to camp on the ruins of New Smyrna, there I noticed more than 400 Houses destroyed, there remained only the chimneys because the Indians, who came each year for the Orange trees which managed to exist in spite of the annual fires, destroyed the wood of which these houses were built to warm themselves.[27]

The 25 Thermometer at 5 Degrees: white frost. I visited the humid places and the surroundings of this Establishment which had flourished in the time of the English; but I didn't notice any other plants besides the ones that had interested me the preceding days. We were now 75 Miles from St. Augustine.[28]

The 26 our navigation was of 12 Miles and we stopped on the ruins of a Plantation which had belonged to captain Besy in a place that was very fertile and which made me want to visit the Swamps.[29]

There I found only one species of Pancratium and an annual Plant 12 feet in height which had dried out and from which I collected several seeds.[30]

The 27 we still navigated between the Isles of Mangroves (Rhizophora Mangle) and we dined at the foot of a hill named Mont Tucker. I collected several shrubs and plants of the Tropics. The evening we came to camp on the ruins of the habitation of Captain Roger.[31]

The 28. I crossed the swamp which once formed this habitation on which Sugar Cane had been cultivated and finally towards midday, we came to the Indian river and for some the Aisa hatcha, that is to say, river of the Deers and called by the Spaniards Rio d'Ais. This habitation was the most southern one that the English had established in Florida. We camped 4 miles further.[32]

The 29 March. Our navigation was around six miles because the contrary wind was very strong, the oarsmen in spite of much effort were not making much progress. Moreover my son and I went on the west bank to try to discover the most narrow place between the Indian river and the Canal where we were. Around 11 from above the trees one could distinguish easily the two Arms of the Sea; that is to say, that one where we were called by the English . . . and the Indian River thus named by the English, which is not a River at all but a very narrow Arm of the Sea like all of the others made so by a chain of Islands which extend from North to South from Carolina to the Cape of Florida. Our two oarsmen came ashore and we walked around the whole territory so as to try to find a less difficult passage to transport the canoe. Around four o'clock in the evening we came back to camp with the hope of being able to transport the canoe. We desired all the more to be near dry ground because since our Departure from New Smyrna we had only brackish water. The provisions of Rum for the oarsmen had been consumed and they were just as eager as we were to leave this place where we were being devoured by Mosquitoes. As for me the place presented alternatively only considerable stretches of joncs and saw palmettos with saw teeth (Chamaerops monosperma fronde acute dentatis radice repente).[33]

However, I found among the trees that made up a part of these Woods situated on the Indian River, a fig tree of entire and oblong leaves, a new Sophora and two other shrubs that were unknown. This increased my hopes for the expeditions that I prepared to undertake in the following days on this River.[34]

Sunday 30 March we have been occupied all day rolling our canoe on the ground, the space of one Mile across rushes and spiny plants. It was necessary to cut trees, but the greatest difficulty was when we had to cross an area of 100 toises[35] all covered with Chamaerops with saw teeth that not only cut our Boots and our Legs but resisted by the strength of their stems the good instruments with which we were furnished. In effect, a very skillful worker whom I had hired for this trip said that he would much rather cut a 60-ft tall cabbage palm than one of these shrubs because the sprawling stem is often interlaced with other stalks or branches of the same size passing one over the other. Finally towards evening, we got the canoe across and all of our baggage transported to the bank of the Indian River.[36]

The 31 March. we were ready to leave at daybreak. But the place where we were was a kind of a Gulf which (to the judgment of our oarsmen) formed with the river a stretch six miles in width. The wind was against us and there was so little water in this part of the Gulf that our canoe could not advance even though my son and I had traversed more than four miles in the water which only came to half way up our legs. When the water became too deep, we would get back in the canoe, but then the Waves came into the canoe and by midday we stopped near a swamp full of Mangroves. Not being able to camp at this place which was very wet mire, we returned to the place from which we had left, but it didn't take long before our canoe was almost submerged because of the large amount of water that came into it and all of our provisions became wet.[37]

Tuesday 1st April 1788 the same wind from the South which had brought us also kept us at the same place. It blew with more violence than the preceding day. Our oarsmen profited from the occasion to dry the Rice and the Biscuits which had been soaked the preceding day. They went fishing and brought back two Fishes which weighed more than 18 pounds each. I looked for plants after having dried my Baggage which had also been submerged the day before and I collected Pteris lineata and Polypodium Scolopendroides which commonly grow together on stems of the large Chamaerops. I found also Acrostichum aureum in very humid places and even among the Mangroves which border the immense swamps of this river. We saw aquatic birds of several species and my son killed that day over 12 by repeatedly firing of his shotgun. We cut Cabbage palms so as to save the bread which was diminishing and we allotted two Biscuits per day for each of the five people.[38]

April 2 we profited from a calm wind by crossing the river to the side of firm Ground. It was at least six miles distant and toward midday we reached shore. The wind which had gotten considerably stronger impeded us from continuing our route after midday. I found on dry Ground an abundance of Sophora occidentalis, a beautiful shrub and I collected an abundance of seeds and a beautiful stalk of its flowers which confirmed that this was a Sophora of which the flower is very attractive. I picked up several other plants which the night impeded me from describing . . . A new species of Spigelia, and another plant which has an affinity with . . .

Our walk was judged to be twelve Miles long.[39]

The 3 April our walk was fifteen miles and the hope of finding interesting and new plants inspired me to overcome the obstacles, because I was still traveling by foot to relieve the oarsmen who had a contrary wind, instead I only found the trees and shrubs of Georgia and of Carolina Magnolia glauca, Gordonia, Acer rubrum. However I collected two Annonas, one of which was a new species with very large white flowers and leaves * * * The length of the Canal, which was 4 to 8 miles wide in several places, frightened our oarsmen and they judged that it was more convenient to profit from the wind to return, so we decided that we could profit from the calm wind which comes everyday just before sunrise until 9 o'clock in the morning. In effect on the 4th having embarked before sunrise with a favorable wind, we had the good luck to traverse the deepest part before 8 o'clock and in the evening we found ourselves on the eastern bank of the Aisa hatcha river.[40]

Every evening we saw from our camp the fires that the Indians made on the other bank of this river, but since our departure from St. Augustine, we had not encountered any directly; our oarsmen advised us to avoid any encounter with them because of the demands to which one is exposed on their part in order to get Rum for which they are at least as passionate as can be. Our Oarsmen, who by the way, were among the most sober I have yet seen.[41]

Our navigation was estimated to be 24 miles.[42]

The 5th was entirely used up in transporting the canoe and rolling it in the same way as we had done on the preceding Sunday.

Toward the evening I profited from a small time interval to collect several shrubs and trees which I had noticed on the shore of this River which I had not seen before. I packed them in such a way as to transport them to Charleston so I could plant them there and everything was ready to return to St. Augustine the next day.

Sunday 6 April before leaving this most Southern part of Florida to which I had been able to advance, I decided to visit an Island where I saw trees different from those (other than only the mangroves) found commonly on these Islands and I did not waste any time picking up the Guilandina bonduccella, the Mangrove with fruits like those of the Catesby fig tree . . .

An unknown tree and a Phaesol or a Dolichos with large fruits.

Our navigation was of . . . and we came to camp on the ruins of the habitation of capt. Roger. This habitation was the most southern that the English had had in Florida. There they had cultivated sugar, but the Indians have destroyed all of the Canes.[43]

The 7, the wind had blown from the south for several days and was very favorable for our return and pushed us all the way to New Smyrna of which there is nothing there but Ruins as I have already remarked. Our navigation was of . . .

The 8 we came to sleep on an Island ten miles distant of . . . We were at the latitude of . . .[44]

The 9. we had the wind behind us and despite our many stops we covered 24 Miles.

We came to camp at the mouth of the Tomoka Creek, latitude of . . .

The 10 We went up the Tomoka river which is truly a River, although it was named a Creek by the English who didn't know Florida very well during the time they possessed it. The Wind was very favorable and we found toward the evening an island covered by woods. We camped a little past the island and our navigation was around 18 miles at best.

I collected an Annona with large, white flowers which I believe to be the Annona palustris and Annona glabra which seems to me to be a variety of triloba. The productions on this river are: Acer rubrum, Cupr. disticha, Fraxinus . . . Magnolia grandiflora and glauca, Pinus foliis binis.[45]

The 11 we went up around five miles but the river which was filled with trees prevented the Canoe from passing so I decided to eat here and to look for plants while the meal was being prepared and to leave immediately after.

That evening we came back to camp at the Mouth of the Tomoka river.[46]

The 12 a man left to go look for some horses so we could transport the Baggage which could not be transported in the Canoe, in order to cross the Matanzas Inlet.

Sunday 13 April the man whom I had sent to the home of the Minorcan to get horses, arrived in the evening and he brought the provisions that we needed.

I employed the preceding day and this one in visiting the surrounding Woods and the swamps where I was, but there were no interesting plants in this disagreeable place because Caimans and Serpents were abundant and Mosquitoes tormented us and didn't let us rest during the night.

The 14 we started walking at daybreak and didn't arrive until very late in the evening due to the detours that we were forced to take several times across saw-toothed Chamaerops which covered the surface of the ground, because the woods are very open. We were required to take a considerable detour because the woods had been on fire the preceding days. They were still burning and the wind brought the fire toward us at a rapid pace. One has no idea in Europe of the considerable amount of woods that are annually set ablaze in America either by the Indians or by the Americans themselves. They have no other motive one or the other than to have new grass come up without the dry grass of the preceding year. I am persuaded this is the principal cause of the decline of the North American forests, and this burning only allows people to hunt Deer more easily and to feed the Cattle that roam in the woods.[47]

The 15 we waited for the oarsmen who had gone by Sea to cross Matanzas Inlet.

I went to look for plants in the woods and I noticed the Andromeda which I had seen before as being truly a new species, having enough resemblance with the Andromeda arborea, but different in several regards particularly by the disposition of its flowers and . . .

I also recognized an Andona and Stillingia silvatica. I collected all the rare shrubs and trees to finish a box which I had decided to take to Charleston, at some risk since the season was then very advanced.[48]

The 16 we left this place to return to St. Augustine and we came to camp two miles from Ft. Matanzas.

The 17 we set out at two o'clock in the Morning and we arrived in St. Augustine (the wind having been favorable) at Midday.

The 18 I went to visit the Spanish Governor and I visited Mr. Leslie, agent for Indian affairs and to discuss with him the means of traveling in Indian land.[49]

The 19 I was invited to dinner by Mr. Leslie.

Sunday 20 April I received a visit from the Governor who came to see my Plants and other Collections that I had made during my trip, birds etc. I was invited to have dinner with him and the afternoon was spent in the gardens of His Excellency with the hospitable Ladies of his family.[50]

The 21. 22 & 23 I looked for plants in the area around St. Augustine and I sent a man to the St. Johns River to reserve a canoe to shorten the trip by keeping us from entering the river at its mouth.

The 24. 25 & 26 I wrote Monsieur le Comte d'Angivill [d'Angiviller] to give him an account of my trip to the south, my Collections and to announce to him the draft of 2000 francs to the order of M. De la Forest drawn on M. Dutartre.

Wrote to M. l'Abbé Nolin to respond to his letter received here and to make some observations on the Plants that I am sending him.

Furthermore I asked him for the racine de Disette and for seeds of the male Veronique for Mr. captain Howard. I have written to M. De la Forest to send him the bills of exchange drawn on M. Dutartre in triplicate to employ funds for the establishment near New-York.

I also wrote to M. Dr Marbois, consul of France in Philadelphia, to register the package Addressed to M. le Comte d'Angivill. This week I described several grasses, Carex, Scirpus and other plants which grow in the area around St. Augustine.[51]

Sunday 27 April, wrote out the Lists and the Descriptions of the Plants collected since my arrival numbering 40 species whose genera and species are well known to me.

The second notebook contains 36 whose genera are well known to me, but the species are doubtful or unknown.

And the 3rd Notebook contains 29 the greater part of which are unknown or could not be determined because I have not seen them in flower.

In all 105 Trees or plants collected since the 1st of March until this day.[52]

The 28 April bought provisions and prepared to leave to go visit lake George situated beyond the St. Johns river.

Gave the letters written earlier to captain Hudson who was to leave to go to St. Mary to get his ship and go to New-York making a stopover in Savannah. Wrote at the same time to M. Ferry Dumont.

Addressed the package to Mr De la Forest as well as the bills of exchange on Mr Dutartre.

Observed at the settlement of St. Roquet Annona grandiflora in abundance.[53]

The 29 We left for the St. Johns River.[54]

The 30 we arrived at the residence of Mr. Wigin situated on this river 40 Miles from St Augustine by land.[55]

Thursday 1st May 1788, I looked for plants in the surroundings and I collected an Androm. formosissima in flower. The Canoe being ready the 2nd of May, we embarked and we passed by the Store established for commerce with the Indians situated 10 miles away [from where we started]. We camped further away and we have navigated sixteen miles on this River.[56]

The 3rd May we traveled between 14 and 16 Miles still having contrary wind and we camped in a place called camp des Indiens, which seems to have been cultivated at one time. I recognized Sapindus saponaria, some orange trees, and a pretty Convolvul. dissectus? etc.[57]

Sunday 4 May we traveled only four miles and we camped on an Island at the entrance to lake George on the east bank opposite a place named the point of Alligators. The wind which was contrary and very strong, forced us to stay in this place where I recognized Erythrina, again woody, and Sapindus saponaria. The woods were filled with sour Oranges.[58]

The 5 May, we saw as we entered Lake George a big deep Bay to our left, that is to say, to the West and after having directed our route straight ahead we entered a river that one cannot see before arriving at a distance of only twenty toises. The mouth (at 29 Degrees 5 Latitude) of this river is so filled with sand that it was necessary to drag the Boat the distance of twenty-five to 30 toises. Then one finds an area over 15 feet in depth. The water is salty and more disgusting than that of the St. Johns river and that of lake George. After having gone upstream for more than three miles, we found the spring which comes out the ground and forms Bubbles which rise up over half a foot on the surface. One can see the bottom at greater than 30 feet in depth. Around the Basin formed by this spring, we recognized the Illicium. The ground was composed of sand darkened by plant debris and Shells.

The other trees which abound in this place, as well as everywhere where one finds Illicium are Magnolia grandiflora and glauca, Ilex cassine, Olea amer. and Laurus Borbonia. This river abounds so prodigiously with Fish that they would hit our canoe as we traveled along. Our course was five miles up to the mouth of this River.[59]

The 6 May we continued up river again following the shore and as I was walking on the sand while the canoe continued, I recognized at a Mile away from the place where we had left, that is to say, from the mouth of the salt river, a spring of water, the purest and the best that I had yet drunk in Florida. We stopped there to eat, because we were all thirsty and disgusted with the bad water that we had been drinking for several days. One mile further I again recognized the Illicium and it was found in abundance at the southern point of the Bay.

After having passed the bay, (29 Degrees 3 latitude) we came to camp at the Hill of Oranges to shelter ourselves from a furious Storm which was about to descend on us. At the bottom of this Hill is the Mouth of a fairly wide river the water of which is not as good as the preceding water. I went up this river about two miles and I recognized in the woods Sapindus Saponaria. A species of Coffea which I had observed before at Mosquito shore and two other trees that I had seen there, but which had remained unknown to me. I also saw Crinum americanum. Our journey was judged to be 15 miles long.[60]

7 May 1788, our navigation was eight miles. We passed Lake George and we entered the River which lies above and we camped in a Place abundant with Orange trees. We arrived early enough to construct a Shelter of leaves of the wild Palm Chamaerops . . . so as to be safe from a storm.[61]

8 May, our navigation was 10 Miles and we endured a storm more severe than the one of the preceding day. We saw a Place frequented by Indians. There was a boat that belonged to them on the bank of the river and a cooking pot. I put some Biscuits, beans, and some sweet Oranges in this Pot and we continued on our way. We heard two gun shots which proved that the Indians were hunting nearby. We passed through a place so abundant with oranges that I traveled more than one-half mile through the interior of these Woods without seeing any other trees. This place was over one Mile long. We came to camp on a hill where I recognized Rivina humil., an Asclepias shrub & &—the Gledisia montosperma at the bottom of the hill and the summit was covered with Orange trees.[62]

The 9 May our travel was evaluated at 12 miles only although our Oarsmen had worked all day long, but since our departure with the opposing current, since we were going up river, the wind was always contrary. For more than eight miles, we saw nothing on either side of the river but rushes and a few trees, the ground was always muddy. The river was bordered on both sides by Alligators or Caimans which with their horrible faces were of an enormous length and size. We approached them from 6 to 10 feet away. Their form is that of a

Lizard, but they are black and armed the whole length of their back with big points which they bristle up when they are angry. One cannot kill them but by charging the gun with balls and aiming at the lower Neck. The Nose is more turned up than that of a pig, the head flattened two feet & four inches in length sometimes even longer. The eyes are very close to the top of the head. They have 72 teeth in the Jaw. They easily swallow Dogs, Pigs, and young calves, but at the least movement of a man, they jump in the water with a big crash. They are amphibious and came every morning to visit us to get the leftovers of the Fish with which we were well furnished on this river. We were also regaled with their Music the sound of which resembles a stronger and more continuous Snore than the Bellowing of a Bull, situated in a valley one mile away. The Indians eat the lower part of the alligator sometimes, but only when they lack other game.[63]

10 May our navigation was 15 miles; we continued up to the source of a river which came out of the ground. The water was salty and gave off an unbearable odor, although one could see the bottom of the river at more than 15 to 20 feet in depth. We had much difficulty passing over trees that covered the bottom and sometimes obstructed the surface. There were no habitations more remote from the time of the English than these ruins where we ate lunch. I found at the most secluded place to which we had advanced a species of wild squash.[64]

Sunday eleven May, we traveled 11 Miles still continuing to go upstream against the current of the River which seemed more and more obstructed and losing itself in Marshes covered with rushes. I picked up an Ipomoea, the flower of which was perfectly white and the tube six inches long. This plant appears to me to be an annual and grows in moist places, the leaves are entire, cordiform. Seeing little success in continuing my Voyage, I made the group turn back and we returned to sleep at the same place from where we had left the same day.[65]

12 May, the wind was favorable for our return and we traveled twenty-seven Miles. We camped at the Hill of Orange trees.[66]

13 May, the Wind and the Current were again very favorable & we arrived at the shore of the stream whose water was so agreeable and beautiful. It is situated only one-half a mile from the salt water river [Salt River] the water of which is just as bad as the water of the little river is good. I experienced furthermore the satisfaction of collecting at only eighty toises distant the Illicium. It should be remarked that

this shrub is found in places where the Magnolia grandiflora, the Annona grandiflora, Olea americana, Ilex cassine etc. etc. grow but more particularly where one also finds Aralia spinosa and a Grass called "Canes" which grows to ten feet in height which always indicates a good but sandy and cool soil. Our trip for the day was 18 to 20 Miles.[67]

14 May our navigation was of * * * and we arrived at the home of Sr Wigins . . .[68]

15 May we started back on the route by land to return to St. Augustine.[69]

16 May, we arrived at St. Augustine at two o'clock in the afternoon.

The 17 I went to visit his Excellency the Governor etc.[70]

The Sunday 18 May, I wrote down my collections.

19 I was invited to dine at captain Howard's house.[71]

20 and 21 I went to look for plants at the extremity of the Isle of St. Anastasia.

The 22 was Corpus Christi and I went to the Procession.[72]

The 23 took leave of His Excellency the Governor & several other people of distinction from whom I had received a favorable reception.

The 24 gave the Government a detail of the observations made in Florida during my stay.[73]

The Sunday 25 May left St Augustine for the Poste of St. Vincent and we slept at Twenty-Miles house.[74]

The 26, our horses having strayed during the night, we searched for them the following morning. The Sergeant of this Poste who was in charge of our horses had us led by two Soldiers and two other horses to the Poste of St. Vincent situated 40 miles from St Augustine.[75]

The 27 we embarked in our canoe which had come by Sea to wait for us at Poste St. Vincent because we had already taken advantage of a small ship which had set sail for this part of Florida.[76]

28 May 1788, we navigated between Islands of rushes and we camped opposite la Barre de Nassau river.[77]

29 May, we arrived at the mouth of the St. Mary river which separates Florida from Georgia and we camped in the territory of Georgia.[78]

Commentary

As early as April 2, 1787, Michaux had considered visiting Florida. A letter of April 8, 1787, written from Charleston to Count d'Angiviller indicates that the botanist was ready to explore Georgia and even go to Florida if he did not experience difficulties with the Spanish or Indians (appendix 2-7). His *Journal* account shows that he and François André, now 17, left Charleston on April 19. On May 6th of that year, he, his son, and a servant were at Sunbury, Georgia, the principal town and seaport on the barrier island of St. Catherines. The thriving shipping port was second only to Savannah in importance (Vanstory 1981). William Bartram wrote of the prosperous port town consisting of several two-story houses. Today, the port of Sunbury is gone without a trace.

The explorers attempted different ways to travel to St. Augustine, but without success; the Florida visit had to be postponed. On that same day, François André, the servant, and an unnamed English traveler, later identified in the *Journal* as John Fraser, traveled to the banks of the Altamaha River. André senior remained at a shelter six miles from Sunbury because of a leg injury caused by an insect sting, which had become inflamed and abscessed from constant friction against his horse while riding. The next two days André botanized in the Sunbury area. On May 10th the party headed back to Augusta and later went to the southern Appalachians with Indian guides.

Michaux's plans to return in the winter to Georgia and to the country of the Cherokee to collect seeds failed. Unrest prevailed among the Creek, Cherokee, and Chickasaw against the Georgians and Carolinians. In the autumn of 1787 the hostile activities between the contenders culminated in a war. Michaux wrote to Count d'Angiviller from Charleston on February 9, 1788, to tell him of the demise of the Georgia trip (appendix 2-9).

In late January of 1787, Governor Zéspedes made a four-week tour of his province to assess the Spanish defenses and resources of the land. His visit to the northeast section of Florida included areas along the St. Marys and St. Johns Rivers. In the governor's party were Captain Carlos Howard, Father Miguel O'Reilly, Chief Engineer Marino de la Rocque, Zéspedes's son Lieutenant Vizente Domingo de Zéspedes, and military personnel (Zéspedes 1787; Tanner 1963).

As the year progressed, tension and unrest increased for the Spanish governor. He feared that the Indian-Georgian war might extend to East Florida. Rumors were that the British were trying to obtain control of the Indian fur trade throughout eastern North America. Other rumors were that Spain might turn Florida over to France. By December of 1787, restlessness had increased among the inhabitants of the St. Marys River area. A group of men wanted Zéspedes to replace his border agent, former British officer Henry O'Neill, whose duties were to prevent illegal trading across the Florida-Georgia border (Tanner 1963).

The disturbing times of 1787 continued for Zéspedes into 1788. Because of the Georgia border crises and other troubling events that arose in 1787, Zéspedes formulated a new defense plan for his province in January 1788 (Tanner 1963). In late March, Daniel McGirth, a thief who had been ordered to the Bahamas in 1786, suddenly appeared in the St. Marys River area and was arrested by Henry O'Neill. McGirth had received his passport in December 1787 from the governor of the Bahamas. Then in April, William Augustus Bowles, whose motive was to take East Florida from the Spanish with the help of Indians, landed in Florida. Bowles was a former member of the Maryland loyalist regiment and husband of a daughter of Perryman, the Lower Creek leader (Tanner 1963). Until his death in December 1805 at Morro Castle at Havana, Bowles continued to create problems for Panton, Leslie and Company, Zéspedes himself, his successor, and American officials (Siebert 1929). To add to Zéspedes's headaches, border agent O'Neill was mysteriously murdered in May of 1788; his assailant was never identified.

Notes

1. Although Michaux omitted from his *Journal* and other writings any descriptions and impressions of St. Augustine and the immediate area surrounding the town, one can surmise with confidence, by gleanings from historical records, what the visitors saw. As the ship approached St. Augustine, they would have seen the north end of Anastasia Island lying directly east of the town (map 4). Thirteen days after arriving in St. Augustine the explorers would be botanizing on the island and visiting its most famous inhabitant, Jesse Fish.

One of the first, if not the first, man-made structures André Michaux's eyes perceived as the party entered the Port of St. Augustine was the Castillo de San Marcos, formerly called Fort St. Marks by the British (photo 10; map 14). This stalwart fortress, constructed entirely of coquina blocks quarried from Anastasia Island, rests near the edge of the Matanzas River at the north end of the town. The formidable-looking, diamond-shaped landmark of medieval architecture, with its four bastions and protective moat, had been, and still is, a fixture of the town since construction began in October 1672. The fort was finished in 1695.

Photo 10. Castillo de San Marcos, St. Johns County, Florida. An ancient landmark seen by André Michaux in 1788, the fort can be visited and enjoyed today.

Figure 2. City gate at the northern end of St. George Street, St. Augustine, St. Johns County, Florida.

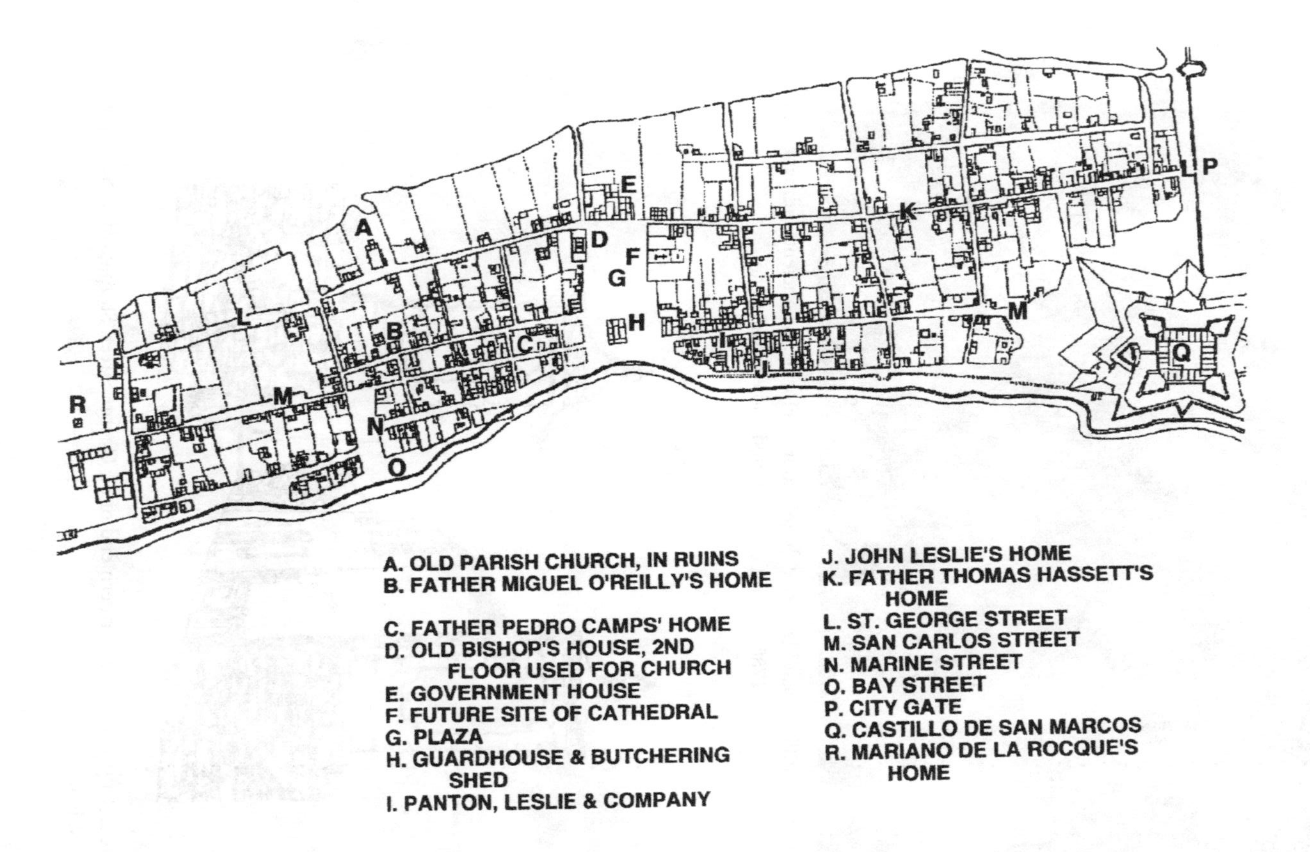

Map 14. St. Augustine, based on Mariano de la Rocque's map of April 25, 1788 (redrawn by Richard Jackson; modified from Tanner 1963).

Photo 11. St. Augustine, St. Johns County, Florida, street scene. Photo courtesy of Florida State Archives.

Except for some destroyed and dilapidated houses, St. Augustine was a typical Spanish town whose physical appearance had changed little in two hundred years (Johnson 1989). The gridiron or somewhat parallelogram layout, about three-quarters of a mile long and less than half a mile wide, was situated on a peninsula. The three main streets ran irregularly north and south, parallel to the Matanzas River. These were intersected by several east-west streets. The city gate, which today has changed little with time, was at the north end of St. George Street (figure 2; map 14). The unpaved streets of sand, hardened by shells and lime, were especially narrow; sidewalks were nonexistent (photo 11). Most dwellings were built of coquina rock, two-story, and with windows projecting like balconies into the streets. Windows on the north side of the houses were lacking because of the harsher exposure that prevailed in the winter. Some of the houses had chimneys that were built by the British (Schoepf 1911). Soils of the town and immediate surroundings were sandy, marshy, and not very fertile (Dewhurst 1885).

Nearly four years had passed since Zéspedes had assumed the governorship. Based on a Spanish census taken by Father Hassett in September 1788, the population of East Spanish Florida was 1,729 (1,078 whites and 651 Negroes), exclusive of the military garrison at Castillo de San Marcos (Mowat 1943; Tanner 1963). This number is higher than the 1,390 individuals (900 whites, 490 Negroes) Zéspedes calculated during his successful trip of January–February 1787 (Zéspedes 1787). For the city of St. Augustine, Zéspedes gave the total population to be 568 Spaniards, Minorcans, Italians, Greeks, Islanders, Floridanos, and royalists and 169 blacks. These numbers excluded troops or employees of the king.

2. The day that Michaux arrived in the Port of St. Augustine, he, while on board the ship, was approached and questioned by officers of the Spanish government. They asked him his purpose for coming to St. Augustine, and wondered if he had brought with him any merchandise. Michaux responded with his intentions to study the natural history of Florida, and mentioned that permission to visit East Florida had already been made. Nonetheless, the botanist was told that he had to personally appear before Governor Zéspedes. Michaux told the governor that he would give him a report of his most interesting discoveries while in Florida.

Because of the recent turbulent events and recently formulated defense plan of January 1788, the personal interview with Zéspedes was probably a precautionary measure. That Michaux had contacted the Spanish government prior to his visit is recorded in his *Journal*, and is also indicated in a letter to Count d'Angiviller dated February 9, 1788, from Charleston: "I am hastening to take advantage of the permission of the Spanish governor to go to Florida" (appendix 2-9).

A letter to Zéspedes requesting permission to visit Florida before André Michaux left Charleston has not been located, nor has a letter been found from Zéspedes granting Michaux permission to explore East Florida. On the other hand, in the East Florida papers (East Florida Papers 51) deposited in the Library of Congress, with copies at the P. K. Yonge Library of the University of Florida, an unpublished letter to Zéspedes exists with André Michaux's signature in which Michaux requests permission to explore Florida; however, the letter, written in Spanish from St. Augustine, is dated March 8, 1788 (photo 12 and appendix 2-18). Why such a letter was written on that date and from St. Augustine cannot be explained. In this letter Michaux refers to himself as Professor of Botanical Sciences. This is strange, but shows that Michaux was not above aggrandizing his position to give himself more status. He never used such a title elsewhere nor did he ever have

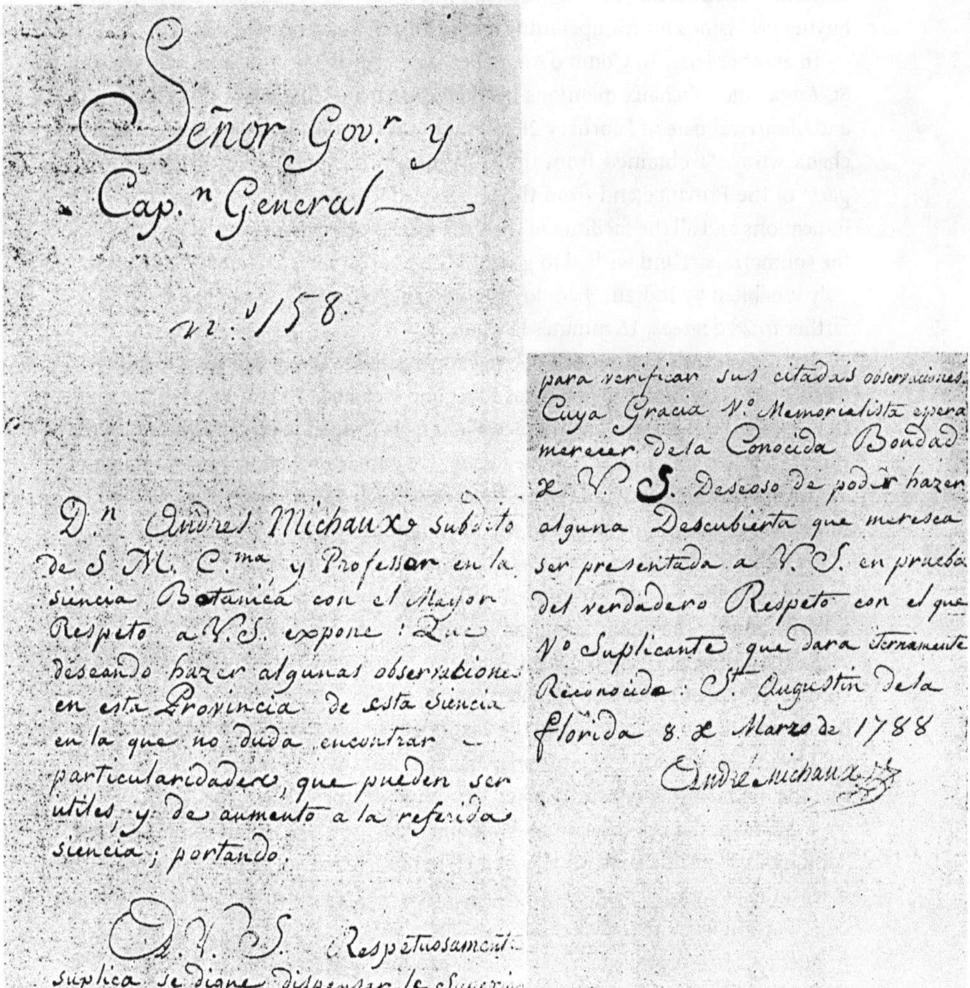

Señor Gov.r y
Cap.n General

n.o 158

D.n Andres Michaux subdito de S. M. C.ma y Professor en la siencia Botanica con el Mayor Respeto a V. S. expone: Que deseando hazer algunas observationes en esta Provincia de esta siencia en la que no duda encontrar particularidades, que pueden ser utiles, y de aumento a la referida siencia; portando.

A V. S. Respetuosamente suplica se digne dispensar le Superior permiso, para que pueda libremente viajar x en ella acompañado de dos practicos, y x un Negro x su ...

para verificar sus citadas observaciones. Cuya Gracia V.o Memorialista espera merecer dela Conocida Bondad x V. S. Deseoso de poder hazer alguna Descubierta que meresca ser presentada a V. S. en prueba del verdadero Respeto con el que V.o Suplicante quedara eternamente Reconocida: St. Augustin dela Florida 8. x Marzo de 1788

André Michaux

Photo 12. Letter written by André Michaux to Governor Vizente Manuel Zéspedes requesting permission to explore Spanish East Florida. The letter, in Spanish, is dated March 8, 1788, and sent from St. Augustine. See appendix 2-18 for translation of letter. By permission of Archivo General de Indias.

that title. Based on his *Journal*, Michaux was indeed in St. Augustine on March 8, buying provisions for his upcoming trip to Florida's east coast.

In another letter to Count d'Angiviller, dated April 24, 1788, and written from St. Augustine, Michaux mentions his departure from Charleston on February 14 and his arrival date of February 28 in St. Augustine, Florida (appendix 2-10). Michaux wrote, "I obtained from the Government the permission to travel in all parts of the Province and from the Officers attached to the Government all the indications and all the facilities at their disposal. I directed my initial trip towards the southern part and wished to go until the Cape of Florida, but as this area was only inhabited by Indians, I could only advance to Cape Canaveral and a little bit farther to 28 degrees, 15 minutes latitude."

The Governor's House, also Government House, though in poor condition, was a short walk from the Matanzas River and located at the west end of the town square called the Parade, later named Plaza de la Constitución. The Parade, at the center of town, was a large oblong area where parades were formed and much of the town's business activities were transacted (map 14). Orange trees planted by British Governor Grant bordered the north and south ends of the Parade. Many of the public buildings and homes of the wealthy and prominent individuals were located near the Parade (Johnson 1989). At the east end of the Parade near the waterfront were the public slaughter house and public market.

3. The initial meeting between André Michaux and His Excellency was very cordial. Zéspedes welcomed Michaux, showed him every courtesy, and extended his services to the explorers from the very beginning and until Michaux and party left Florida. He was solicitous toward Michaux and sent word at his place of lodging that the botanist was to be well taken care of while in Florida.

4. Michaux did not delay in making himself known, regardless of his location. Although no specific individual is named in the *Journal*, he probably visited with at least the government officials.

5. For the first time Michaux looked for plants in the surrounding area of St. Augustine. He recognized an *Andromeda* of a new species, No. 1, 2, & 3. Based on IDC 57-11, the third specimen is without question the rusty lyonia or rusty staggerbush, *Lyonia ferruginea* (Walt.) Nutt. On the collection label Michaux wrote, "No 3. *Andromeda* folius revolute subt—ferruginis; St. Augustine." Rusty lyonia is also listed in the *Flora*. Four other *Andromeda* species Michaux collected from "Floride," with no specific locality given, were the coastal plain staggerbush, *Lyonia fruticosa* (Michx.) G. S. Torr. (*Andromeda coriacea*? Michx., IDC 57-1); maleberry, *Lyonia ligustrina* (L.) DC (*Andromeda paniculata* L., IDC 56-13 and 56-16); Piedmont staggerbush, *Lyonia mariana* (L.) D. Don (*Andromeda mariana* L., IDC 57-14); and the swamp doghobble, *Leucothoe racemosa* (L.) A. Gray (*Andromeda racemosa* L., IDC 57-20). Any of the four species named above could represent No. 1 and No. 2.

Located in the herbarium in Paris is a specimen of the sensitive brier, *Mimosa quadrivalvis* L. (*Mimosa horridula* Michx. Environs de St Augustin, *Mimosa vi-lotica*?, IDC 127-13). The plant was collected in the area of St. Augustine, but is not mentioned in the *Journal*. There are several species that occur in the Michaux collection and not in the *Journal*; the reverse condition also applies.

6. Mass was held on Sunday. This is one of the few places where Michaux mentions going to Mass. Because the condition of Roman Catholic properties, including the parochial church, was deplorable, Zéspedes had fitted the second floor of the Bishop's House for worship services (Gannon 1965). The first floor was occupied by a guard room, temporary jail, and storage area (Dewhurst 1885; Tanner 1963; Tebeau 1980). This building, located at the south end of the Parade, had been improved and used by the British as a courthouse (map 14). At the time Michaux visited St. Augustine, the site for the future Roman Catholic cathedral on the northwest end of the Parade had been set aside as shown on the 1788 Rocque map of St. Augustine (map 14). The cathedral, erected during 1793–1797 and constructed of coquina stone, was designed by engineer-architect Mariano de la Rocque. Two Irish priests, Father Thomas Hassett, aged thirty-seven in 1788, and Father Miguel O' Reilly, aged thirty-six, served the Catholic families of St. Augustine. Father Thomas Hassett was vicar-general and the principal religious figure of the province.

7. The Réaumur thermometer was an alcohol thermometer developed by French physicist and naturalist René Antoine Ferchault de Réaumur (1683–1757). The thermometer's scale ranged from 0 degrees (freezing point of water) to 80 degrees (boiling point of water).

8–9. The second botanizing trip was thwarted by a storm, accompanied by thunder and lightning, which continued through the night and into the next day. After the weather subsided the next day, the botanist ventured more than six miles from St. Augustine. Michaux saw the same interesting species of *Andromeda* shrubs (No. 2 and No. 3) as observed on March 1. The exact species of the fourth plant that resembled an *Andromeda*, but was unlike the others in having different fruit, has not been determined.

10. The most likely source for Michaux's description of Florida was the book by Bernard Romans (1775). Later in his *Journal* the botanist mentions the "habitation of Captain Roger," calls the Indian River the Aisa Hatcha, and describes present-day Mosquito Lagoon and Indian River as "Arms of the Sea." These same descriptors for these sites are used by Romans in his book. It is also possible that Zéspedes allowed Michaux to read the account of his January–February 1787 Florida exploration, which was written on April 15, 1787. The map he used is not known, but it could be one from Romans's book or one of William Gerard De Brahm's maps.

11. Michaux does not give us a description of his canoe. However, Deleuze

Figure 3. Sketch of a canoe, the type commonly used during early Florida times. The vessel was a hollowed out trunk of a large cypress tree. André Michaux's canoe was probably similar to the one illustrated.

(1804, 1810), a personal friend of Michaux in France, provided a detailed description of the vessel. It was made from one bald cypress trunk hollowed out along its length and fitted with a sail. The canoe measured about twenty-two feet long, but was only three feet wide and two and one-half feet deep (figure 3). Two persons could not sit side by side. This type of dugout was commonly used for riverine transport throughout the Southeast, including the trip of the Bartrams during their 1765–1766 explorations on the St. Johns River. Deleuze (1804) erred in stating that Michaux obtained the canoe at the Tomoka River.

12. The identity of the civilian Michaux visited on March 11 to make arrangements for a depot for plants he would collect is not known. The next day, Wednesday, five people—Michaux, his son, the two oarsmen, and the male Negro, who came with Michaux from Charleston, boarded the canoe and left St. Augustine for the southern coastal areas of East Florida. The person at whose house on St. Anastasia Island (also called Fish's Island) the party stopped, because of the unfavorable tide, was Jesse Fish, one of the most colorful and oldest inhabitants of the St. Augustine area (map 4).

Michaux was correct in his statement that Fish had lived on the island for fifty-two years. When Spain ceded East Florida to England in 1763, Jesse Fish acquired by deed 185 houses and a larger number of lots in St. Augustine. In addition to the ten thousand acres on Anastasia Island where he lived, Fish owned more than 850 acres in different tracts, most of which were on Moosa Creek (Siebert 1929). Fish claimed that he purchased the land on Anastasia Island from the Spaniards during the First Spanish Period. Fish, originally from New York and an agent of the Walton Exporting Company of that city, came to St. Augustine in 1735. Thus, not only was he a resident during the First Spanish Period, but he also lived under the British government and into the Second Spanish Period (Mowat 1943; Tebeau 1980).

Fish's home on the island, called El Verge (the Garden), was known for its elaborate gardens with olive, date, lemon, and sweet orange trees (Forbes 1821; Mowat 1943). The citrus groves on Anastasia Island were world famous (Siebert 1929). François André Michaux (1804), fifteen years later, reminisced about Fish's beautiful plantation, which he had seen in 1788, and about the quality of the oranges, which he remembered to be very sweet, large, and thin-skinned. These oranges were more esteemed than those of the Caribbean. François indicated that the five hectares of orange trees on Anastasia Island that belonged to Fish originated from seeds that came from India. Fish was the first person to export oranges and a spirituous beverage called "orange shrub" (Tanner 1963). Bernard Romans (1775) stated that Fish's place was more pleasant than profitable. André Michaux called it a paradise—a paradise that had withstood pillages by pirates and survived domination by two countries, England and Spain. Michaux was obviously impressed with Jesse Fish when he described him as being the most hard working and most industrious man in all of Florida. Despite his enterprising achievements, Fish was in debt at the time of his death in February 1790, about two years after Michaux's visit. Fish was buried on his plantation on Anastasia Island, where his gravestone was later destroyed by vandals.

13–14. The mangrove Michaux recognized on the shore of Anastasia Island was the red mangrove, *Rhizophora mangle* L. This species no longer occurs on the island because of past freezes. Red mangrove is not in the *Flora*, but there is a Michaux specimen of this species from Florida in the general collection of the Paris herbarium. The Florida arrowroot or coontie, *Zamia pumila* L. (IDC 125-2; photo 13) still grows on the island. The coontie occurs in Michaux's collection and

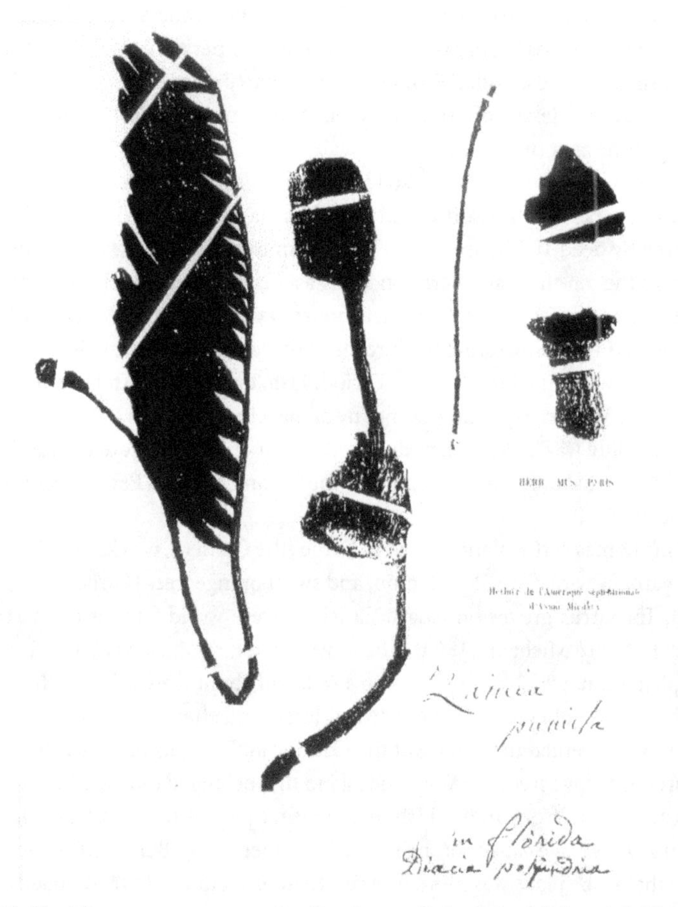

Photo 13. Florida arrowroot or coontie, *Zamia pumila* L., from Michaux's herbarium. André Michaux found this species on Anastasia Island. IDC 125-2. By permission of the Inter Documentation Company, Leiden.

Photo 14. Fort Matanzas, St. Johns County, Florida. An old landmark visited by André Michaux in 1788 in his search for plants.

is mentioned in the *Flora*. In the evening the explorers arrived at Fort Matanzas, situated on the southern end of Anastasia Island (photo 14).

Although Michaux stated that he looked for plants a short distance from the fort, none is specifically listed in the *Journal*. In the Michaux herbarium in Paris, several labeled plants were collected in the area of Fort Matanzas: the grass *Eriochloa michauxii* (Poir.) Hitchc. (*Panicum molle* Michx. Lieux tres humides a 15 miles de St Augustin, IDC 12-11); melicgrass, *Melica mutica* Walt. (*Melica glabra* Michx. No. 5 Florida f Matança, No. 5, IDC 15-5); *Psilotum nudum* (L.) Sw. (*Psilotum floridanum* Michx. No. 45, a 2 Miles du Fort Matança, Floride, sous les orangers, IDC 132-12); and Carolina silverbell, *Halesia carolina* L. (*Halesia parviflora* Michx. circa Matança in the *Flora*; *Halesia* a petit fruit, in Floride, IDC 82-14, photo 15). In the *Journal* Michaux calls the silverbell *Halesia tetraptera*, a synonym for *H. carolina*.

15. The "barre" where the party tried to pass the morning of March 14 is Matanzas Inlet, where Anastasia Island terminates. Because of this problem, Michaux left the canoe with the oarsmen in the area of Ft. Matanzas and he and his son stopped at a Minorcan's house located three miles from the mouth of the Northwest River. Today, the Northwest River is called Pellicer Creek; it borders St. Johns and Flagler Counties (map 4 and photo 16).

16. It is obvious that Michaux and his son were cordially welcomed because they remained at the Minorcan's place for the next five days. Michaux wasted no time in exploring the area for plants. He recognized the same species that he had seen in Carolina and Georgia, including the southern magnolia, *Magnolia*

Photo 15. Silverbell, *Halesia carolina* L., from André Michaux's herbarium. This plant was observed by Michaux near Fort Matanzas and along the Northwest River (Pellicer Creek of today). IDC 82-14.

grandiflora L. (IDC 73-12); willow oak, *Quercus phellos* L.; loblolly pine, *Pinus taeda* L.; wax myrtle, *Myrica cerifera* L.; yellow jessamine, *Gelsemium sempervirens* (L.) Ait. (*Bignonia sempervirens*); and a *Juglans* hickory, probably the water hickory, *Carya aquatica* (F. Michx.) Nutt. All species occur in Michaux's collection and in the Flora.

17. The horse was probably from the Minorcan who was also the guide for the trip up Pellicer Creek (photo 17). Michaux thought the river was misnamed and that it should be called a stream or a creek, as it is today. The trip was by foot,

Photo 16. The Matanzas River looking west at the entrance of Pellicer Creek. At the time of Michaux, Pellicer Creek was called the Northwest River. Photo courtesy of Jerry Millen.

and the horse was probably used to carry their baggage and plants. Again plants common in Georgia and Carolina were listed: magnolia, wax myrtle, loblolly bay; red maple, *Acer rubrum* L.; red bay, *Persea borbonia* (L.) Spreng. (called *Laurus borbonia*); and bald cypress, *Taxodium distichum* (L.) L. C. Rich. (*Cupressus distichia* L.). In addition to these, they found along the river sourwood, *Oxydendrum arboreum* (L.) DC (*Andromeda arborea* L.); coontie; saw palmetto, *Serenoa repens* (W. Bartr.) Small (*Chamaerops repens*); Carolina silver bell with small flowers; two unnamed species of *Asimina* (*Annona*); a leguminous shrub (No. 17); and another unnamed shrub (No. 18). Both the sourwood and silver bell have not been recorded in the area in recent years. The unnamed shrub 18 is today's redroot or littleleaf buckbush, *Ceanothus microphyllus* Michx. (*Rhamnus microphyllus* Michx. le 16 Mars lieux arides vers Nord West river, IDC 37-5; photo 18). This specimen is the type for the species. Plants of *Ceanothus microphyllus* occur today at Faver-Dykes State Park located on Pellicer Creek (photo 19). All identified species mentioned above occur in Michaux's herbarium and *Flora*.

Sargent (1889), in his annotation of Michaux's *Journal* (page 28), suggested that the No. 17 specimen was the coralbean, *Erythrina herbacea* L., probably because Michaux described the leguminous shrub as having ternate (ternées) leaves. The *Erythrina* specimen in the herbarium collection (IDC 86-20) lacks data on locality; however, there is a specimen, but not *Erythrina*, in the Michaux herbarium (IDC

Photo 17. Pellicer Creek looking east where it soon meets the Matanzas River. Michaux, his son, and an unnamed Minorcan traveled on horses along the river.

Photo 19. Redroot or littleleaf buckbush, *Ceanothus microphyllus* Michx. Michaux found this plant on today's Pellicer Creek. Compare this photo with photo 18.

Facing page: Photo 18. Redroot or littleleaf buckbush, *Ceanothus microphyllus* Michx., from André Michaux's herbarium (IDC 37-5). This plant was discovered by André Michaux along the Northwest River (Pellicer Creek of today) and is mentioned in the *Journal* as unidentified plant No. 18. By permission of the Inter Documentation Company, Leiden.

84-19) with a label that contains the note "No. 17 Psoraloides arbuste legumineux Florida, Georgie, Kentucky; Lieux humides pres Nord West riv. en Floride." An annotated label attached to the specimen, but not in Michaux's handwriting, correctly identifies the specimen as *Petalostemum carneum* Michx. Because the No. 17 in the *Journal* record and locality matches the specimen, the plant mentioned in the *Journal* is probably not *Erythrina*, but whitetassels, *Dalea carnea* (Michx.) Poir. or *Petalostemum carneum* Michx. of some botanists. Michaux was familiar with the *Erythrina* in Florida, as can be seen from his *Journal* entry for May 4. Today, both *Dalea* and *Erythrina* occur in St. Johns County, Florida.

18. Michaux collected a lot of seeds from the *Dalea* and from a new *Andromeda*. Other plants observed on the 17th include Walter's viburnum, *Viburnum obovatum* L. (*Viburnum cassinoides* L.; the IDC 41-5 labeled *Viburnum cassinoides* is *V. obovatum* and not *V. nudum*), rattan vine, *Berchemia scandens* (Hill) K. Koch (*Ziziphus scandens* Michx.), and *Lupinus pilosus flore cerulea*. All species, except the lupine, are in the *Flora* and collection. The lupine is mentioned only in the *Journal*. The lupine is probably the sky blue lupine, *Lupinus diffusus* Nutt., that is recognized today as a distinct species from the lady lupine, *L. villosus* Willd. That distinction was not made by Torrey and Gray in 1838–1843 (Torrey and Gray 1969). *Lupinus villosus* appears in the Michaux *Flora* and herbarium as *Lupinus pilosus* (IDC 85-20, 1787 near the Savannah River). *Lupinus diffusus* occurs today in the Florida locality in question, whereas *L. villosus* is less common. The other lupine mentioned by Michaux in the *Flora* and *Journal* is the sundial lupine, *Lupinus perennis* L. (IDC 85-19), obtained in Georgia. Having reached an area of continuous arid soils, probably a scrub habitat, Michaux decided to turn back because he did not find any interesting plants.

19. The *Andromeda*, with leaves like those of the almond, is known today as the leucothoe or coastal doghobble, *Agarista populifolia* (Lam.) Judd (photo 8). The plant occurs in Michaux's *Flora* as *Andromeda laurina* Michx. and in the herbarium collection (IDC 57-10).

20. Because André Michaux had not heard from the two oarsmen and his servant, he returned to Fort Matanzas and found that the oarsmen had already successfully crossed the Matanzas Inlet (photo 20). On the way back to the Minorcan's place, Michaux passed an area full of oranges and two miles further he found several interesting shrubs, but none is specifically identified.

21. After spending five days at the Minorcan's house, Michaux and son set out to join the oarsmen. The Minorcan furnished the explorers with three horses, and, although not specifically written in the *Journal*, the Minorcan went along with them.

22. Michaux probably followed the Kings Road south to the Tomoka River. Michaux called the Tomoka River the Tomoka Creek. The plantation that Michaux

Photo 20. Looking northeast toward Fort Matanzas from a point just north of the Matanzas River and Pellicer Creek junction. Michaux may have walked across this area to the fort, when the land was probably drier.

mentions formerly belonging to Governor John Moultrie was Rosetta (maps 5 and 6). Savage and Savage (1986, page 377) erronously refer to this plantation as Moultrie's elaborate and more famous Bella Vista. Bella Vista was located only four miles south of St. Augustine near Moultrie Creek of today. Camp was on the bank of the lagoon, the Tomoka River Basin of today, and contact was made with the oarsmen who had arrived earlier.

23. Though not mentioned, the Minorcan returned home with the horses. Michaux was about a mile from the mouth of the Tomoka River (photo 21).

24. The abandoned habitation on the left bank of the Lagoon was undoubtedly Mount Oswald (map 5 and 6). This once-thriving plantation was owned by Richard Oswald, a wealthy Scots merchant, West Indian planter, and British envoy at the Paris negotiations in 1782, also attended by Benjamin Franklin. The identity of Orange Island, where camp was made, has not been determined. This site of many orange trees was four miles from the abandoned plantation formerly overseen by James Penman (maps 7 and 8). Several abandoned houses, which could have constituted a village, were noted and visited by Michaux.

Even though no plants are specifically mentioned in the *Journal*, Michaux did collect specimens, now housed in the herbarium at Paris, from the Tomoka River area. All were sedges: *Fimbristylis spadicea* (L.) Vahl (*Scirpus castaneus* Michx. Hab. in Florida juxta Tomoko creek, IDC 8-5); *Rhynchospora ciliaris* (Michx.) C. Mohr (*Schoenus ciliaris* Michx. Hab. in Florida, juxta Tomoko Creek legi, IDC

Photo 21. Mouth of the Tomoka River, Volusia County, Florida. On April 10, 1788, Michaux and party entered the Tomoka River at its mouth.

9-10); and *Fuirena scirpoidea* Michx. (*Fuirena scirpoidea* Michx. le 11 April vers le haut de Tomoka riv. 25; IDC 10-9; photo 22). The latter species was obtained on his return trip to St. Augustine.

25. The papaya, *Carica papaya* L., observed about four miles from the mouth of Spruce Creek, is native to tropical America and was introduced early into Florida where it had escaped from cultivation. William Bartram located the plant in 1774, growing on the banks of the St. Johns River south of Mud Lake. Michaux mentions the papaya only in the *Journal.*

26. Camp was made on Easter Sunday between Ponce de León Inlet (la Barre de New Smyrna) and the ruins of New Smyrna established earlier by Doctor Andrew Turnbull (maps 7–9). Michaux recounted with much accuracy the story of the New Smyrna tragedy when he wrote that more than twelve hundred people, men, women, and children—the greater part of them from the island of Minorca—had been seduced by false promises and led from their country. The descriptions of Turnbull and his colony by Michaux are similar to those of Romans's (1775) account, another indication that Michaux had read Romans's book on Florida's natural history.

27. The Michaux party camped on the ruins of New Smyrna, where the chimneys of more than four hundred houses remained. Apparently the Indians used the wood from the houses for fuel to warm themselves.

28. A white frost in late March does occur in Florida.

29. The plantation in ruins where the Michaux party stopped was Mount Plenty, which formerly belonged to Captain Robert Bisset (maps 7 and 10). The swamp

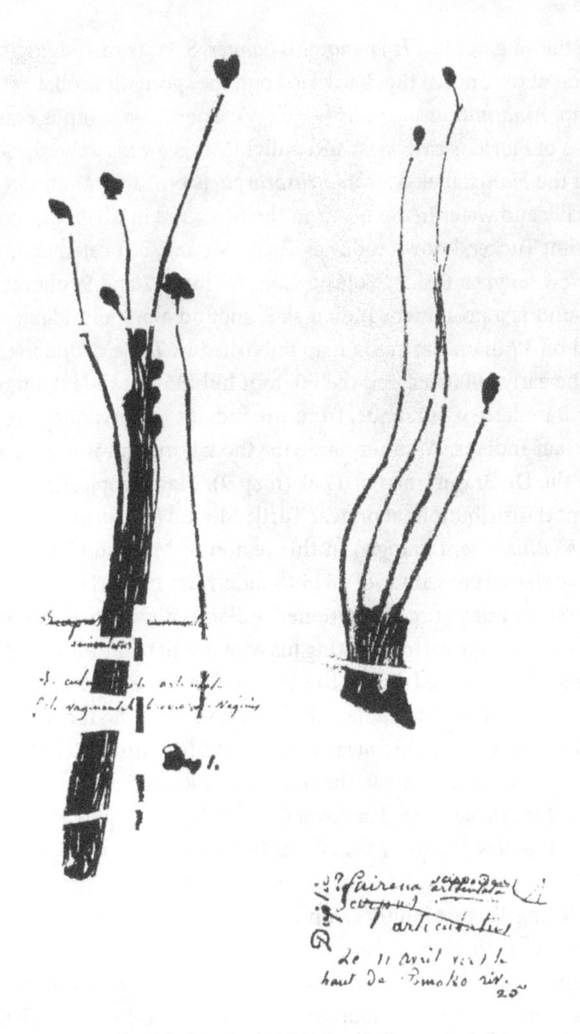

Photo 22. *Fuirena scirpoidea* Michx. André Michaux discovered this species on April 11, 1788, near the Tomoka River (*Fuirena scirpoidea*; le 11 April vers le haut de Tomoka riv.). IDC 10-9. By permission of the Inter Documentation Company, Leiden.

was located west of Mount Plenty. In this general area, Bisset had several tracts with numerous buildings, but none is mentioned by Michaux.

30. The herbarium specimen labeled *Pancratium* (IDC 43-20) has a single flower that looks like the mangrove spiderlily, *Hymenocallis latifolia* (Mill.) M. Roem., commonly found today in the swales, mangrove swamps, and other habitats of Florida's east coast. Leaves from the herbarium specimen are much too

wide to be the alligator lily, *Hymenocallis palmeri* S. Watson. Judging from the size and location of the annual that had dried out, the species is probably the southern water hemp, *Amaranthus australis* (A. Gray) Sauer. This hemp is common today in wet areas of Florida's east coast and individuals grow to twelve feet or more tall. Plate 50 of the *Flora*, labeled *Acnida rusocarpa*, is probably *Amaranthus australis*. The spiderlily and water hemp occur in the *Flora* and in Michaux's collection.

31. Mount Tucker, known today as Turtle Mound, is located about eight miles south of New Smyrna Beach, Volusia County (maps 7 and 9; photos 23 and 24). Turtle Mound is a prominent Indian shell mound along Florida's east coast and is featured on William Bartram's map published in *Travels*. One map of the area made in the early 1600s named the 60–foot hill (35 feet today), largely of oyster shells, the Baradero of Surruque, from the Indians of Surruque, a subdivision of the Timucuan Indians. Another name for the tall mound was Mount Belvedere, shown on the De Brahm map of 1771 (map 9). Many tropical plants reach their northernmost distributions at or near Turtle Mound (Norman 1976). The lush red mangroves (*Rhizophora mangle*) in this region of Mosquito Lagoon were killed back during the severe winter of 1835 (Sunderman 1953). Dr. Jacob Rhett Motte (1811–1868), an army surgeon stationed at Fort Ann, wrote in his journal of the persistent dead mangrove trees during his winter visit to the area in 1837. Fort Ann was a military post erected during the Second Seminole Indian War (1835–1842). Map 15 shows the fort located next to the east bank of the Indian River and near the old Haulover Canal, which was dug in 1854 (Griffin and Miller 1978), much after Fort Ann was abandoned. The old canal is located about seven-eighths of a mile south of the present-day Haulover Canal bridge at Highway 3, Brevard County, Florida. The new Haulover Canal was in operation in 1887–1888 (Griffin and Miller 1978).

After leaving Turtle Mound, camp was made in the evening on the ruins of the plantation of Captain Roger, also spelled Rogers (map 5). In Michaux's *Journal* entry for the next day, he, like Romans (1775), says that the Roger habitation, a sugarcane plantation, was the most southern plantation that the English had established in Florida. Both Romans and Michaux evidently erred in naming Captain Roger as the owner of this most southern plot of land. The southernmost plantation during the British time belonged to Peter Elliot, a British East Florida absentee owner (D. Schafer, pers. comm., June 2000) (map 11). John Ross, a Scot, was one of several managers for Peter Elliot. Probably Romans, and then Michaux, confused Ross with Roger.

32. The swampy area where Michaux crossed, probably on foot, was undoubtedly the southern extension of the Great Swamp that was located on or near the Bisset plantation. The swamp formed the head of the Indian River. Both Aisa Hatcha and Rio d'Ais were mentioned by Romans (1775) for the Indian River. Camp

Photo 23. Turtle Mound, Volusia County, Florida. This shell mound was called Mt. Tucker by André Michaux. Mt. Belvedere is another name that has been used for this prominent feature on early Florida maps. The mound, largely composed of oyster shells, was built by Timucuan Indians. Many Florida tropical plants reach their northern limit at Turtle Mound. André Michaux dined at the base of the mound and remarked on finding several tropical plants there. Photo courtesy of Todd Campbell.

Photo 24. Mosquito Lagoon looking north from Turtle Mound, Volusia County, Florida.

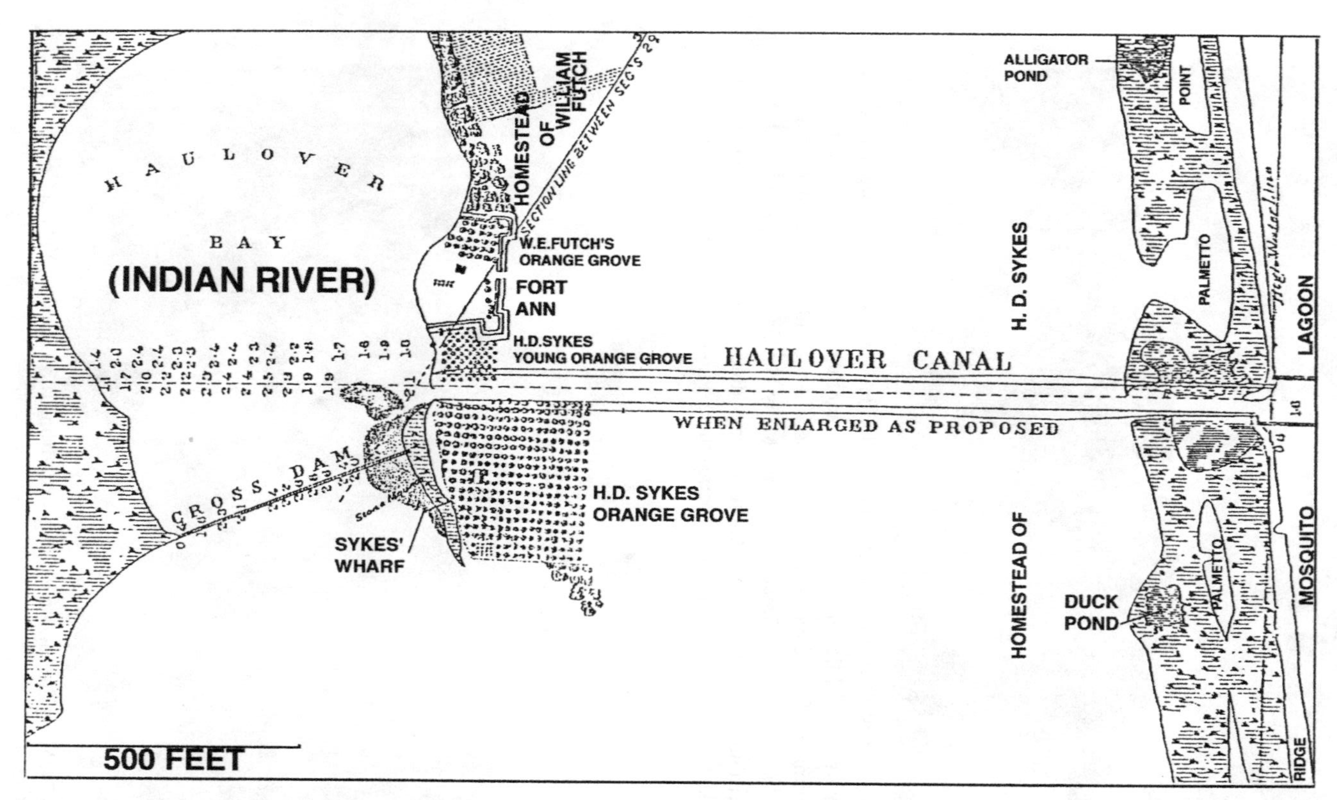

Map 15. Map showing the proposed old Haulover Canal, which was dug in 1853 and 1854, and Fort Ann, present in 1837. Brevard County, Florida (modified from Gillmore 1883).

for the 28th was probably along Mosquito Lagoon. In 1837, Dr. Jacob Rhett Motte described the southern part of Mosquito Lagoon as "an immense extent of flat land, overflowed by the sea, giving it the appearance of a long lake, the water nowhere exceeding six feet in depth."

33. The canal is present-day Mosquito Lagoon. Michaux must have climbed a tree to see both Mosquito Lagoon and the Indian River. From the elevated Haulover Canal bridge of today, a great view can be obtained of both bodies of water. The narrowest place between the Indian River and Mosquito Lagoon is the area known as the Crossing Place or Haulover (maps 10 and 11).

Early maps of this area, such as those of William Gerard De Brahm's 1771 and Mexia's map of 1605, given in Higgs (1951), show a "Crossing Place" or "Haulover" where Indians dragged or "rolled" their canoes over felled trees across the narrow strip of land between Mosquito Lagoon and the Indian River. Present-day Haulover Canal, a man-made canal that connects the two bodies of water, is about seven-eighths of a mile northwest of the older canal that was built in 1854 (map 15). Even today, the hordes of salt water mosquitoes are a loathsome nuisance. One can appreciate the desires of the explorers to leave Mosquito Lagoon. The continuous stretch of "joncs" probably consisted primarily of black needlerush, *Juncus roemerianus* Scheele. Today in many bodies of water along the east Florida coast there are thick stands of the rush. The *Chamaerops monosperma* is saw palmetto.

34. The discovery of a strangler fig, *Ficus aurea* Nutt., IDC 106-10, a new *Sophora*, and two unknown shrubs in the woods along the east bank of the Indian River increased Michaux's hopes of finding more novelties in the days ahead. What species the *Sophora* represents is not known with certainty, but it is probably not the yellow necklacepod, *Sophora tomentosa* L. that Michaux found four days later in flower.

35. A toise is 1.949 meters.

36. The Chamaerops with saw teeth is saw palmetto. The tall cabbage palm *Sabal palmetto* (Walt.) Lodd. ex Schult. & Schult. f. occurs in the *Flora* and in the Michaux collection. A one-mile passage through scrub and saw palmetto where they dragged or rolled the canoe is about the correct distance across the land following the old Haulover Canal from Mosquito Lagoon to the east bank of the Indian River. In 1837, Dr. Motte in his journal described the narrow strip of land between Mosquito Lagoon and Indian River as "one unbroken expanse of scrub-saw-palmetto from three to four feet high" (Sunderman 1953). Motte wrote that the Haulover at his time consisted of a rude path left by Indians that connected Mosquito Lagoon with the Indian River.

After spending nearly all day, Michaux and his workers got the canoe and baggage transported to the east bank of the Indian River.

37. Today, the widest part of the Indian River is near the old Haulover Canal and former site of Fort Ann (map 15). Dr. Motte wrote in 1837 in his journal that the width of this part of the Indian River adjacent to Fort Ann was six miles in breadth and that the body of water was shallow (Sunderman 1953).

38. All of the ferns that Michaux lists in the *Journal* are still common along the east Florida coast: shoestring fern, *Vittaria lineata* (L.) J. Smith (*Pteris lineata* Michx. sur les Bords de le Riv Aisahatcha le 1 Avril, Floride, IDC 129-10; in the *Flora*, the plant is named *Vittaria angustifrons* Michx. Hab. in Florida, juxta amnem Aisa-hatcha) and swamp fern, *Blechnum serrulatum* L. C. Rich. (*Polypodium scolopendroides*), which grows on the stems and branches of cabbage palms. This last fern is not in the IDC collection, but certainly the species is *B. serrulatum* from Florida, which is in the *Flora* and in the Jussieu herbarium with the catalog number of 1388A (Morton 1967). In the *Flora*, Michaux specifies the habitat as "in Florida, juxta amnem Aisa-hatcha," which is the Indian River. Michaux also found that day the giant leather fern, *Acrostichum danaeifolium* Langsd. & Fisch. (*Acrostichum aureum* L. sur la riv. Aisa hatcha Floride, IDC 130- 18), growing in the wet areas and among mangroves that border the immense swamp of the Indian River (photo 25). At the time of Michaux *A. danaeifolium* was not distinguished from *A. aureum* (Morton 1967). The central part of the growing end of cabbage palms is still eaten today. The giant leather fern occurs in the *Flora* and in the Michaux collection.

39. The explorers crossed the Indian River to the west bank. There growing on dry ground was an abundance of the yellow necklacepod, *Sophora tomentosa* L. (*Sophora occidentalis* L.), a species still common in this area along the east Florida coast. Four days earlier Michaux had secured a plant he called a new *Sophora*. No mention was made whether this plant was in flower. Sargent (1889) assumed that this first plant was *Sophora tomentosa* L. Because the second plant was blooming and because Michaux wrote the scientific name of the plant in the *Journal*, he clearly knew the necklacepod. *Sophora tomentosa* L. is not found in the Michaux herbarium or in the *Flora*. Michaux did name other plants of *Sophora* and today these plants are in the genus *Baptisia*. Therefore, it is probable that the nonblooming "Sophora" was a *Baptisia*.

No species of *Spigelia* occurs today in the area where Michaux traveled in Florida. The new species of *Spigelia* mentioned in the *Journal* is the water pimpernel, *Samolus ebracteatus* Kunth, which is in the *Flora* and in the Michaux herbarium (IDC 27-6, *Samolus* [crossed out on the label] *Lysimachia* affinis: No. 8, sur la riv. Aisa hatcha en Floride le 2 Avril 1787) [1788 sic]. This plant is mounted next to the pineland pimpernel, *Samolus valerandi* L. (*Samolus parviflorus* Raf.), whose description is on the same piece of paper bearing the description of *Samolus ebracteatus*. *Samolus valerandi* could be the "unknown plant" mentioned in the *Journal*. Both species of *Samolus* occur today in the area of concern.

Photo 25. Giant leather fern, *Acrostichum danaeifolium* Langsd. & Fisch, from Michaux's herbarium. The fern was collected by André Michaux on the Indian River (Sur la riv. Aisa hatch Floride). IDC 130-18. He called the fern *Acrostichum aureum* L. At Michaux's time *A. danaeifolium* had not been distinguished as a separate species from *A. aureum*. By permission of the Inter Documentation Company, Leiden.

40. Again, Michaux wrote of finding trees and shrubs previously observed in Georgia and Carolina: the sweetbay, *Magnolia virginiana* L. (*M. glauca* L.), loblolly bay, and red maple. The new species of *Asimina* (*Annona*) that Michaux describes as having very large white flowers was probably the bigflower pawpaw, *Asimina obovata* (Willd.) Nash (*Orchidocarpum grandiflorum* Michx. IDC 73-22; photo

26). All identified species are in the *Flora* and in the Michaux collection. The species of the other *Asimina* mentioned has not been determined.

41. Although Michaux did not encounter any Indians during the entire trip, he decided not to venture to the Cape of Florida as originally planned because of possible Indian conflicts.

42. It is difficult to determine exactly the route Michaux traveled in the Merritt Island-Cape Canaveral area. He spent five full days here looking for plants. In a letter written and dated April 24, 1788, from St. Augustine, Michaux told Count

Photo 26. Flag or bigflower pawpaw, *Asimina obovata* (Willd.) Nash (*Annona grandiflora* Bartr.). This beautiful pawpaw was found by Michaux on the Indian River and at the residence of Mariano de la Rocque, chief engineer at St. Augustine.

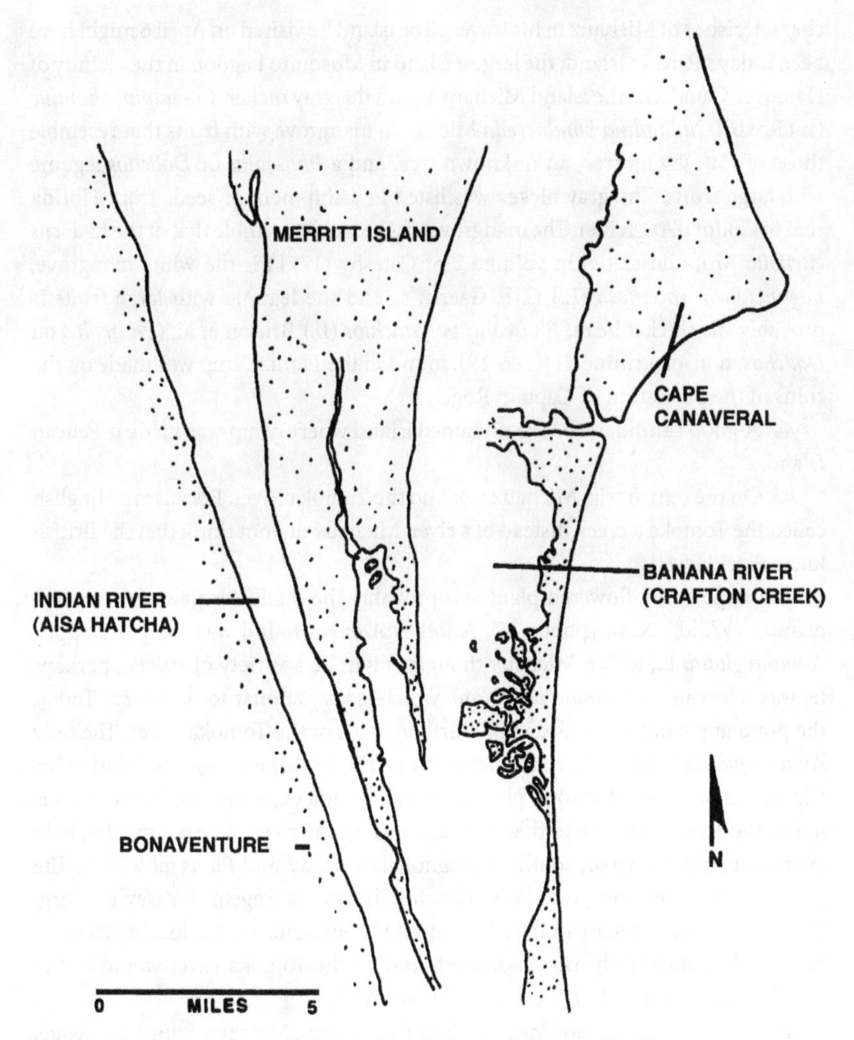

Map 16. Southern section of Merritt Island, Brevard County, Florida.

d'Angiviller that he went a little past latitude 28 degrees 15 minutes. Judging from the mileage Michaux gave in his *Journal* along with the latitude mentioned, the explorers probably went just past the present-day town of Bonaventure, Brevard County, Florida (map 16). This is probably further south than William Bartram traveled along the east Florida coast in the winter of 1766.

43. For the return trip, most areas previously traversed were revisited. The pace of travel was quicker and the written accounts in the *Journal* become briefer, two

characteristics of Michaux in his travels. The island he visited on April 6 might have been today's Pardon Island, the largest island in Mosquito Lagoon in the vicinity of Haulover Canal. On the island Michaux found the gray nicker, *Caesalpinia bonduc* (L.) Roxb. (*Guilandina bonduccella* Michx.), a mangrove with fruits that resemble those of Catesby fig tree, an unknown tree, and a *Phaseolus* or *Dolichos* legume with large fruits. The gray nicker was listed in a shipment of seeds from Florida sent to Count d'Angiviller. The mangrove whose fruits resemble that of the fig *Ficus citrifolia* Mill. illustrated in volume 2 of Catesby (1771) is the white mangrove, *Laguncularia racemosa* (L.) C. F. Gaertn f., and the legume with large fruits is probably the thicket bean, *Phaseolus polystachios* (L.) Britton et al. (*Phaseolus* ou *Dolichos*, non-determiné, IDC 86-19), found in the *Flora*. Camp was made on the ruins of the plantation of Captain Roger.

44. A good candidate for the unnamed island where camp was made is Pelican Island.

45. On the return trip, Michaux went up the Tomoka River. Because the English called the Tomoka a creek instead of a river, Michaux did not think that the British knew Florida well.

The large white-flowered plant was probably the bigflower pawpaw, *Asimina obovata* (Willd.) Nash (photo 26). Another *Annona* found was the pond apple, *Annona glabra* L., which Michaux thought might be a variety of *triloba*; perhaps he was referring to *Asimina parviflora*, which is very similar to *A. triloba*. Today, the pond apple occurs considerably further south of the Tomoka River. The only *Annona glabra* L. in the Michaux herbarium (IDC 73-18) was collected during his trip to the Bahamas that took place after the Florida explorations. The species is not in the *Flora*. Other trees that Michaux saw on the river were red maples, bald cypress, an ash, *Fraxinus*, southern magnolia, sweetbay, and *Pinus follis binis*. The pine is surely the sand pine, *Pinus clausa* (Chapm. ex Engelm.) Vasey ex. Sarg. Three species of ashes are in the *Flora* and IDC, but none has the locality listed as Florida. The most likely ash candidate found on the Tomoka River would be the Carolina ash or pop ash, *Fraxinus caroliniana* Mill.

46. Although he did not mention it in the *Journal*, Michaux found the sedge, *Fuirena scirpoidea* Michx. (*Fuirena scirpoidea* Michx. Le 11 April vers le haut de Tomoka riv., IDC 10-9) on this date (photo 22).

47. Michaux's dislike for alligators and snakes is evident here and elsewhere in his *Journal*. He was also distraught with the annual burning practice of the Indians and others to stimulate new growth of grasses. We know today that controlled burning is healthy for Florida's fire-based communities.

48. The identity of both the *Andromeda* and *Annona* [Andona, sic] are not known. The former had been seen before and resembled *Andromeda arborea*, but differed in the arrangement of its flowers. Queen's-delight, *Stillingia sylvatica* L.

(*Stillingia silvatica*. IDC 119-15) is fairly common in the area. The species is in the *Flora* and in the Michaux herbarium.

49. Mr. Leslie is John Leslie of Panton, Leslie and Company. Michaux was mistaken in saying that Leslie was an Indian agent. As previously mentioned, Leslie was one of the owners of the company that operated the Upper and Lower Stores on the St. Johns River and traded merchandise with the Indians. John Leslie's house was located on the waterfront (map 14).

50. It is clear from this account as well as others that a cordial relationship between Zéspedes and Michaux had developed.

51. While waiting the return of the man sent to the St. Johns River, Michaux caught up with his letter writing. Letters were written to Count d'Angiviller; Monsieur Dutartre, treasurer of the king's buildings; Antoine René Charles Mathurin de la Forest (1756–1846), French Consul General in New York; l'Abbé Nolin; and François Barbe de Marbois (1745–1837), secretary of the French Legation in Philadelphia. In the letter to l'Abbé Nolin, Michaux asked for "root of the disette" and "seeds of the male veronique" for Captain Carlos Howard. Professor Paul Hiepko (pers. comm., March 30, 1998; June 3, 1999) of the Berlin Botanical Garden and Botanical Museum indicated to us that the disette was one of the varieties of the beet, *Beta vulgaris*, and that the male veronique was probably *Veronica officinalis*, a medicinal plant. Several grasses, sedges, and other plants from the area around St. Augustine were described.

Although not mentioned in the *Journal*, Michaux also planned to send to St. Augustine a male specimen of the date palm (*Phoenix dactylifera* L.) so that the lone female palm there could bear fruit (Deleuze 1804).

52. The notebooks that contained the listing and descriptions of 105 species of plants collected since March 1 have not been found.

53. At the home of engineer-architect Mariano de la Rocque, called by Michaux St. Roquet, the botanist found the bigflower pawpaw, *Asimina obovata* (Willd.) Nash (*Annona grandiflora* Bartr.) in abundance (photo 26).

54. The same party of five left St. Augustine for the St. Johns River.

55. The next day they arrived at the residence of Job Wiggins. Based on the Testimentaria (No. 14, 1797) of the death of Jacobo Wiggins, better known as "Job," and a map (date, author unknown) housed at St. Augustine Historical Society and drawn during the governorship of British Governor James Grant, Wiggins lived on the east bank of the St. Johns River (map 1). His property was north of the Plaza of Rollestown (East Florida Papers 58). Rollestown, also called Charlotta and Mount Pleasant, was about midway between present-day East Palatka and San Mateo in Putnam County. On November 10, 1790, Wiggins obtained twelve hundred acres from Governor Juan Nepomuceno de Quesada, who became the next governor after Zéspedes. This was the same tract where he had been residing prior to 1790.

By June 1796, Job Wiggins was dead. He had had seven children, six living at the time of his death, and at least seventeen slaves (East Florida Papers 58; Historical Records Survey 1941). Job Wiggins is the person whom William Bartram, in his *Travels*, called the "old trader" and "old interpreter . . . often my traveling companion, friend and benefactor." Wiggins accompanied Bartram on the trip to the Alachua Savanna, today's Paynes Prairie, where Bartram served as a delegate to the Indians at Cuscowilla, Florida (Harper 1958). Whether Michaux knew about Job Wiggins from his contact with William Bartram is not known. It is clear that one could obtain a canoe from Wiggins, since Michaux had had a man go to Wiggins's place and reserve a vessel. When William Bartram left Florida he gave his canoe, his bark, to Job Wiggins. One wonders if the canoe that Michaux obtained twelve years later from Wiggins might have been the same one that once belonged to William Bartram.

The trip to Job Wiggins's residence on the St. Johns River was overland and about forty miles from St. Augustine, the distance given by Michaux in his *Journal*. Early maps of east Florida, including the one mentioned above housed at St. Augustine Historical Society and the De Brahm maps of 1765 and 1766, show an Indian trail to Mount Pleasant, connecting St. Augustine and Rollestown (map 1). The trail somewhat parallels today's Highway 207.

56. A flowering specimen of the coastal doghobble, *Agarista populifolia* (Lam.) Judd, was collected (photo 8). Nonflowering specimens of this plant had been found on March 17 on Pellicer Creek and on the Matanzas River. Although not stated in the *Journal*, Michaux found the prairie wedgescale, *Sphenopholis obtusata* (Michx.) Scribn. (*Aira obtusata* Michx., IDC 15-4, "in Florida juxta domum Wiggin") at Job Wiggins's place. This specimen represents the type for the species and Job Wiggins's residence is the type locality (photos 27 and 28).

The "store established for commerce with the Savages situated 10 miles from where we left" was Spalding's Lower Store (map 3). The store, a frequent stopping place used by William Bartram, was on the west side of the St. Johns at present-day Stokes Landing, southwest of Palatka, Putnam County. Camp was probably made near today's Welaka Spring.

57. The "Indian Camp," a site that appeared to Michaux to have been cultivated at one time, was probably Mount Royal, near Fruitland, Marion County. This sand mound was named by John Bartram on December 28, 1765, when he and William made their trip up the St. Johns River (photo 29; maps 1 and 3). Mount Royal was located about three miles south of Mount Hope, the shell mound, near Beecher Point. Bartram also named this mound. The Earl of Egmont, former Lord of the Admiralty, and several partners had been granted this property on the St. Johns River in 1768. It did not turn out to be a profitable venture (Schafer 1982). Although Mount Royal is extant, Mount Hope was destroyed by state officials and

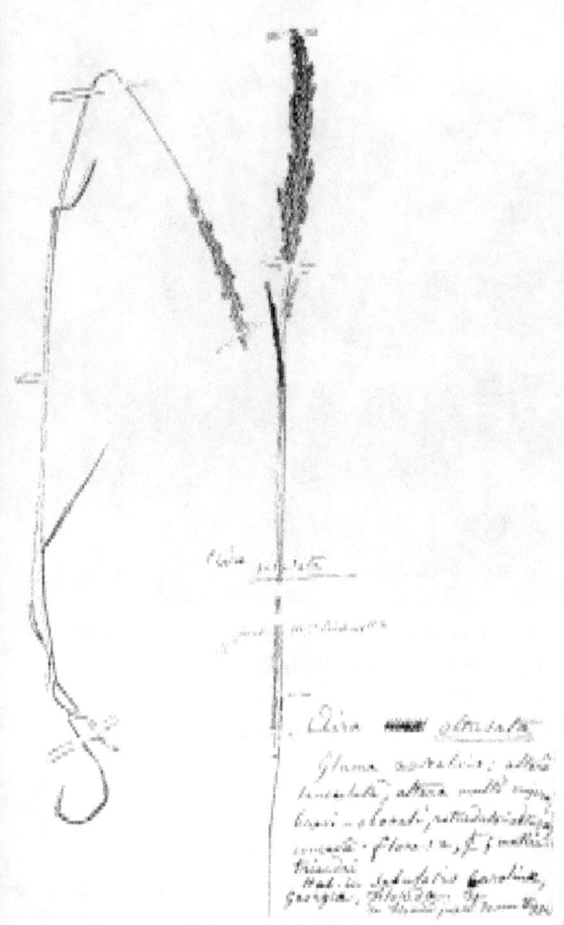

Above: Photo 27. Wedgescale, *Sphenopholis obtustata* (Michx.) Scribn., from Michaux's herbarium. André Michaux discovered this grass at Job Wiggins's residence on the St. Johns River. Michaux called the species *Aira obtusata* Michx. IDC 15-4. By permission of the Inter Documentation Company, Leiden.

Right: Photo 28. Wedgescale, *Sphenopholis obtustata* (Michx.) Scribn.

Photo 29. Mount Royal, a sand mound built by Indians and located near Fruitland, Putnam County, Florida. The mound was named by John Bartram when he and William explored the St. Johns River in 1766. William Bartram visited the mound again in 1774, as revealed in his famous *Travels*. André Michaux camped at Mount Royal in 1788.

its shells used for road construction. When William Bartram visited Mount Royal in 1774, much of the surrounding natural area had been cleared and planted with indigo, cotton, and corn. At the Mount Royal camp, Michaux recognized orange trees, the soapberry, *Sapindus saponaria* L. (*Sapindus saponaria* L., no locality given, IDC 54-19), the Alamo vine, *Merremia dissecta* (Jacq.) Hallier f. (No. 6. *Convolvulus dissectus*?, Floridae, IDC 32-19), and other unlisted plants. The soapberry and vine are in the Michaux collection and *Flora*. William Bartram's first encounter with the morning-glory was in 1774 on Drayton Island (named Kingsley on map 1), about four miles from where Michaux located the plant.

58. The island was Drayton Island and "Alligator Point" evidently is Lake George Point of today. Why Michaux used the descriptive Alligator Point is an enigma. Neither Bartram used that name, and despite extensive search, no other reference was found for Alligator Point on the St. Johns River. On Drayton Island Michaux recognized the coralbean (*Erythrina*) and soapberry, and noted that the woods were filled with sour orange trees. He did not mention the presence of the scarlet rosemallow, *Hibiscus coccineus* Walter, gingerbush, *Pavonia spinifex* (L.) Cav., and

lantana, *Lantana camara* L., all of which were in abundance when William Bartram visited the location.

59. The explorers entered Lake George and proceeded to the west bank. The river that could not be seen until reaching the outfall is today's Salt Springs Run (map 3). The shallow condition of the run near the lake was mentioned by John Bartram on January 24, 1766. He wrote that the creek (Salt River) was three, four, and five feet deep up to the springs, but at the bar of the river the depth was about eighteen inches. That there were numerous fish in the Salt River was observed by John Bartram in 1766 and by William Bartram in 1774.

Michaux saw the famous "crystal fountain" of William Bartram. In 1774 Bartram wrote that the fountain exceeded two to three feet in height. Michaux said that the spring bubbled up to over one-half foot. Surrounding the basin were plants of the rare yellow anisetree, *Illicium parviflorum* Michx. ex Vent. (*Illicium parviflorum*; Floride, IDC 73-3; photo 9). William Bartram was the first naturalist to find the pleasant-smelling yellow anisetree near the spring on January 24, 1766, when he and his father made their first trip to Florida beginning in the fall of 1765 and continuing into early 1766 (Harper 1942). Other trees Michaux noted associated with the *Illicium* were southern magnolia, sweetbay, dahoon holly, *Ilex cassine* L. (IDC 122-8), wild olive, *Osmanthus americanus* (L.) Benth. & Hook. f. ex A. Gray (*Olea americana*), and red bay. The *Illicium*, dahoon holly, and wild olive are in the *Flora* and Michaux collection.

60. The "southern point of the Bay" is Lisk Point (map 3). The "Hill of Oranges" (Colline des Oranges) where camp was made was a shell mound; the fairly wide river is known today as Silver Glen Spring (map 3). On January 23, 1766, John and William Bartram were at this site. John named the site William's Spring. Even though thousands of trees had been destroyed by the time of Wyman's (1875) writing, he said that the orange grove on this shell mound was the most extensive of any on the St. Johns River.

In the woods at Silver Glen Spring, Michaux saw the soapberry and wild coffee, *Psychotria nervosa* Sw. (*Coffea*), which had been observed at Mosquito Shore. Two unknown trees, previously seen at Mosquito Shore, and the string-lily, *Crinum americanum* L., were observed. Both the wild coffee and string-lily occur only in the *Journal*.

61. Although Michaux gave their navigation to be eight miles, the exact place of their camp is uncertain. A good candidate is Zinder Point, a shelly bluff that was covered with orange trees and located near the junction of Lake George and the St. Johns River (photo 30; Harper 1942).

62. The place frequented by Indians was probably the site of Spalding's Upper Store (map 3). Michaux did not mention whether a building existed at the spot. Probably Bluffton or Orange Bluff is the "place so abundant with oranges that I

Photo 30. Lake George looking toward the south entrance into the St. Johns River. Photo courtesy of Boyd Z. Thompson.

traveled more than one-half a mile through the interior of these woods without seeing any other trees." Bluffton was a large shell mound covering about seventeen acres, a short distance below the mouth of Lake Dexter and on the east bank of the river. On January 1, 1766, John Bartram described Bluffton as a "high shelly bluff, with thousands of orange trees" (Harper 1942). The hill where camp was made was probably Bartram's Mound, also called Little Orange Mound or Idlewild Dock. The shell mound, built on the edge of a swamp, was located about two miles from Bluffton on the west side of the St. Johns River. Michaux found at the base of the hill the water locust, *Gleditsia aquatica* Marshall (*Gleditsia monosperma*), and mentioned also seeing the rougeplant, *Rivina humilis* L., and an *Asclepias* shrub (arbrisseau). The water locust is listed in the *Flora* and herbarium; however, the rougeplant is given only in the *Journal*.

63. Michaux did not distinguish the alligator from the caiman. The detailed description of the alligator is similar to that in William Bartram's *Travels*. Both Bartram and Michaux encountered large numbers of alligators in the Lake Dexter area of the St. Johns, and Bartram named the place Battle Lagoon. One wonders from what source Michaux took his detailed information, especially that not readily available from firsthand observation.

64. The source of the river that came out of the ground is Blue Spring, on the east side of the St. Johns River, in Orange City, Volusia County. Michaux was headed

toward Blue Spring (map 3). The salty water, terrible odor, clearness of the bottom, and trees obstructing the passage were all commented on by John Bartram on January 4, 1766, 22 years earlier.

The remote habitation where the party ate lunch was the ruins of the plantation house that belonged to Lord Beresford (map 3). The site is located about four miles from Blue Spring. The Beresford Plantation was near present-day Beresford Station, located on the east side of Lake Beresford toward the north end. At this most secluded place Michaux found a species of squash (coloquite). This squash is probably the rare Okeechobee gourd, *Cucurbita okeechobeensis* (Small) L. H. Bailey. John Bartram in 1766 also mentioned "a native gourd or squash, that ran 20 feet up the trees, close to the river" (Harper 1942). André Michaux and the Bartrams' sightings are probably of the same species. John and William Bartram also found the gourd near High Bluff and about eight miles north of Blue Spring (Harper 1958). All above locations for the gourd, including Michaux's record, are in fairly close proximity to each other. More recently the rare gourd was rediscovered near Hontoon Island State Park by Marc Minno (pers. comm., 1999). The Okeechobee gourd is not in the Michaux collection nor in the *Flora*.

65. After traveling eleven miles and still continuing against the current of the St. Johns River, the travelers decided to turn around. The morning-glory Michaux found is the moonflower, *Ipomoea alba* L. (*Ipomoea bona nox* L., IDC 33-17), a species found in the collection and *Flora*. William Bartram in 1774 found the species near the site where the Okeechobee gourd grew (Harper 1943; Slaughter 1996b).

At this point Michaux ended his voyage up the St. Johns River because he had little success in finding new plants. Based on a straight-line distance of Michaux's eleven miles from Blue Spring, the explorers would have reached High Banks, Volusia County, before returning. Had he continued about five additional miles, he would have entered Lake Monroe.

66. On May 12, the wind and current were favorable and they traveled twenty-seven miles. Camp was made the next day at the Indian Mound at Silver Glen (Hill of Orange Trees).

67. The *Illicium* was collected once more in the area of Salt Springs Run. Michaux commented that this shrub grew where there were southern magnolia, the bigflower pawpaw, wild olive, dahoon holly, especially Devil's walkingstick, *Aralia spinosa* L., and a grass called "canes" (*Arundinaria gigantea* Chapm.) that grew ten feet tall. The cane is in the *Flora* and Michaux collection.

68. The party was back at the residence of Job Wiggins, where their horses were picked up for the return trip. Beginning on May 14 and until May 25 the daily entries in the *Journal* are brief and limited to single sentences.

69. For the return trip to St. Augustine the same overland route was used.

70. Here again after returning from his explorations, Michaux visited Zéspedes.

71. Captain Howard with whom Michaux dined was Captain Carlos Howard, Zéspedes's main assistant.

72. The fête de Dieu was Corpus Christi and Michaux attended the procession.

73. The detailed account of Michaux's Florida observations that he gave to the Spanish government before leaving has not been found, despite much effort in searching the records. That Michaux respected Governor Zéspedes and appreciated his services are indicated in his published *Flora Boreali-Americana* 2 (page 70) where the genus *Lespedeza* was named in his honor: "D. Lespedez, gubernator Floridae, erga me peregrinatorem officiosissimus." The discrepancy between what should have been *Zespedeza* and *Lespedeza* has been attributed to an error made by the printer of the *Flora* (Ricker 1934). However, Hochreutiner (1934) disagrees with Ricker (1934) and contends that the name was not a printer's error, but that Michaux deliberately altered the spelling. Regardless of the discrepancy and controversy, it is clear from Michaux's letter to Governor Zéspedes that he had planned, before the Florida exploration even began, to name a new species of plant in his honor (appendix 2-18).

74. The "Poste of St. Vincent" was known as San Vicente Ferrer. This was a military outpost located forty miles from St. Augustine that was used to guard incoming approaches from the sea. The post was situated on high ground, twelve miles up the St. Johns River on the river's south bank. San Vicente Ferrer was called St. Johns Bluff during the British occupation of Florida. Twenty Mile Post was known as Twenty Miles House and was located twenty miles north of St. Augustine. It was a place where travelers changed horses and rested for the night.

75. There were numerous times during the North American travels that Michaux lost his horses or had them stolen.

76. Michaux had shipped the canoe from St. Augustine to San Vincente Ferrer.

77. Camp was made for the night opposite the bar of the Nassau River.

78. The following morning, they arrived at the mouth of the St. Marys River and camp for the night was made in Georgia.

On his way back to Charleston, Michaux visited the barrier island of Cumberland and navigated along Jekyll Island and St. Simons Island. On Sunday, June 1, Michaux left camp at 2:00 A.M. and arrived on St. Simons at Fredericktown (Frederica), Georgia, at 10:00 A.M. Letters were given to unnamed individuals. That evening he and his son had dinner with James Spalding and ladies of General Lachlan McIntosh's family and several other important people.

Spalding was born in 1734, in Perthshire, Scotland, and emigrated to Charleston, South Carolina, in 1760. He married Margery McIntosh, granddaughter of his business partner, Donald McKay. Spalding lived on St. Simons Island, Georgia, at Frederica, and later at the West Point plantation just north of Frederica (Vanstory

1981). At the time of Michaux's visit Spalding probably resided on the plantation, since many of the houses at Frederica were in ruins even at the time when William Bartram visited the town in 1774 and Spalding was president of St. Simons (Vanstory 1981).

When Spain ceded Florida to England, Spalding established trading posts in Florida. The Upper and Lower Stores on the St. Johns River were the most famous. After the Revolutionary War, Spalding moved to St. Simons Island and died in Savannah on November 10, 1794 (Harper 1958).

At Savannah, Michaux changed their means of transportation and took a ship (navire) back to Charleston, where they arrived on June 8. The next day the botanist and son were at the pépinière. In a letter dated June 10, 1788, from Charleston, Michaux wrote to d'Angiviller that the Florida trip was a success and a very happy one for him because of the large number of rare and new plants obtained as well as the favorable welcome and support that he had received from the governor and other inhabitants of Spanish East Florida (appendix 2-11).

On June 12–14, Michaux planted the cuttings and set out the plants that had been brought back from Florida. During the next two days, the seeds collected from Florida were sown along with a large quantity of other ones collected elsewhere. On July 1, 1788, Michaux wrote to both l'Abbé Nolin, via Mr. Leyritz, and Count d'Angiviller, and sent the latter seeds from plants obtained in Florida (appendix 2-12). The last batch of seeds collected during the Florida trip, along with some seeds from the Carolinas, was mailed and listed in letters dated August 2, 1788, and January 5, 1789 (appendixes 2-13 and 2-16). Seeds were sent from the following Florida plants: *Guilandina bonduncella*, *Caesalpinia bonduc*(L.) Roxb.; *Sophora occidentalis*, *Sophora tomentosa* L.; *Chiococca racemosa* (frutex ericaefolia), *Chiococca alba* (L.) Hitchcock; *Ceanothus (floridanus)*, *Ceanothus microphyllus* Michx.; *Conocarpus racemosa*, *Conocarpus erectus* L.; *Coffea*? nova species, *Psychotria nervosa* Swartz; Hippomane, *Hippomane*; *Amorpha*, probably *A. fruticosa* L.; *Amyris elemifera* L.; and an unknown shrub (arbrisseau). We are uncertain if the *Hippomane* mentioned is the manchineel, *H. mancinella* L., because of the species' very southerly distribution today. Also, in another listing where *Hippomane* is given, Michaux said that the plant was herbaceous. In the letter of January 5, 1789, seeds from the following Florida plants were sent to d'Angiviller: *Sophora occidentalis*, *Sophora tomentosa* L.; an unknown tree, an unknown shrub with opposite leaves; an unknown climbing shrub; *Guilandina bonduncella*, *Caesalpinia bonduc* (L.) Roxb.; *Dolichos* shrub, *Hippomane*? herbaceous; *Zamia pumila* L.; *Ceanothus* new species; *Cacalia*? shrub; unknown shrub bearing berries and having leaves of *Erica*; *Eriocaulon*; *Conocarpus racemosa*, *Conocarpus erectus* L.; *Tillandsia lingulata*, *Tillandsia urtriculata* L.; *Malva caroliniana*, *Modiola caroliniana* (L.) G. Don; *Bombax gossyspinum*; *Chiococca racemosa*, *Chiococca alba* (L.) Hitchcock; *Coffea*?,

Psychotria nervosa Swartz; and *Erythrina arborea*, *Erythrina herbacea* L. This list is very similar to the one sent to d'Angiviller on August 2, 1788.

On October 21, Michaux sent a box to a Captain Marshall so that Marshall could bring him some trees from St. Augustine. Then, on November 1, 1788, Michaux harvested the seeds of *Bignonia sempervirens* (*Gelsemium sempervirens* (L.) W. T. Ait.) and covered the shrubs brought from Florida to protect them from the winter freeze.

From the time Michaux left Florida in 1788, and until his final departure from Charleston to return to France on August 13, 1796, the French botanist's drive to fulfill his assigned mission never waned. This dedication is even more remarkable in view of his constant struggle to obtain past wages owed to him by his government, the return of François André to France in the spring of 1790, the disruptive French revolution that resulted in the deaths of King Louis XVI and Queen Marie Antoinette in 1793, and the loss of his supervisor, Count d'Angiviller, who fled to Germany in 1791 and then to Russia, where he was associated with the Court of Empress Catherine II. D'Angiviller returned to Germany where he died in a monastery (Robbins and Howson 1958).

In spite of orders received in 1790 from Paris to reduce expenditures and cease traveling, Michaux was determined to complete his mission (Michaux 1793). Major explorations made in the eight-year period following his Florida visit in 1788 included trips to the mountains of northeast Georgia and southwest South Carolina (1788), Bahamas (1789), North Carolina mountains (1789, 1794), states of Maryland, Pennsylvania, and New York (1789, 1792), Georgia coast and Cumberland Island (1791), Canadian wilderness to the Hudson Bay area (1792), and to the Old Northwest Territory and Mississippi River, his last and longest trip in America (1795). Letters containing lists of seeds and plants collected from some of these trips are in appendixes 2-14, 2-15, 2-16, and 2-17.

André Michaux's last exploration, which took him to the Mississippi River and the states of Missouri, Illinois, Kentucky, and Tennessee totaled 1,123 miles. The eleven years spent on American soil were eventful and much had been accomplished; however, it was time for the Frenchman to return home.

Epilogue

10th Notebook (Cahier). 1796

JOURNAL

This epilogue follows the same format as given in the previous chapter—the translated account of Michaux's *Journal* precedes the Commentary.

The 27 Thermidor of the 4th year of the French Republic, One and Indivisible (13 August 1796 old style) embarked from the harbor of Charleston, South Carolina on board the ship Ophir [with] Captain Johnston destined for Amsterdam. The 14th and 15th stayed at anchor.[1]

The 16 (30 Thermidor) lifted anchor, set sail.

The 18 lost sight of land.

The 15 (30 fructidor) September Storm which lasted until 16th at night.

5 October passed in the middle of a English Squadron commanded by Admiral Roger Curtis made up of 14 war Boats: 8 Boats with 2 Decks, 2 with 3 Decks, and 4 Frigates. One of the Frigates the Melpomene came in our direction and having sent an Officer on board our Ship, he looked at the Papers and bill of lading of the Captain. Having verified that all the Information given by the Captain was true, he was satisfied and wished him a good trip. In conversation, he said that the war with France was getting annoying to sailors, that they were not taking over any boats but hoped that the war with Spain would be more advantageous and that the first expedition would be against Manille [Manila]. This Squadron was at the entrance of la Manche closer to the Isles of Scilly than Ouessant.[2]

The 18 Vendemiaire in the 5th year of the French Republic, One and Indivisible (9 October 1796 old style) the wind was favorable and good, but at 5 P.M., there arose a Storm which became extremely severe in less than two hours, it continued throughout the night doubling its furor and the wind that came from the east forced us towards shore. At midnight the Captain had prepared the hatchets to cut the Masts. Finally daylight arrived without the boat arriving, but the 19 Vendemiaire, (10th of October) around 8 o'clock the Captain seeing that the Sounding Line did not show an adequate depth of water determined to run the Ship aground; after 4 or 5 violent jolts it stopped; then the waves moved with such rage and violence that everything on Deck was swept overboard. The sails were in shreds in less than 15 minutes. A Mast was broken, the Ship was half overturned and was shaken severely for one-half hour. Then the waves became more violent and we became deluged so that all the crew and myself also were losing their strength. Several Trunks having been brought on Deck were thrown into the sea and the inhabitants of a town named Egmond [Egmont], approximately one league away were removing everything that came near shore.[3] There were 200 people of whom 25 men belonged to a troop with an Officer sent to try to rescue us if possible. Finally several Seamen, having lost hope, attached themselves to pieces of wood which had been thrown overboard and reached the shore. As for myself I had attached myself to some rigging and had my legs under a spar which had become detached during the night and became attached to the Deck. Having been beaten by the Waves for three hours, I felt my strength ebbing and I went down to the lower deck and waited for the end of my suffering and death.[4] Immediately I lost consciousness as I remember nothing of the circumstances which passed until the moment when I was transported to the village, I was undressed and my clothes were changed. They made me drink two little glasses of wine and they brought me near a big fire, where I regained consciousness half an hour later; but I was shaking during the whole day. I only know what was told to me, since I had lost consciousness, that around eleven the Captain having seen the Canoe fall at the base of the Ship got three men who had stayed behind, to transport me in this Canoe as well as another man who was in the same situation. Then they put me in a cart to transport me to the village and towards one P.M. I became conscious.

I was in a Room near a big fire with new clothes and about 40 or 50 inhabitants of the Area. I thought immediately of my Boxes and Trunks which contained my Collections of which I had seen several thrown in the Sea three hours before. The people said that everything which fell off the ship or had been thrown overboard was arriving on shore and that the detachment of troops was on the look out to keep the People from taking anything.

The Captain who was the last to stay behind on the ship threw himself in the water and swam to shore, around two o'clock, and arrived in a Cart in the Village as he was extremely exhausted as well as were all the sailors.

The inhabitants of the Area furnished us all the help possible, Shirts, Suits, Bread, Meat, Brandy etc. and towards evening all the People who had been shipwrecked were relieved and recovered.

The total of my Collections was made up of sixteen Cases and four Trunks of which only 5 or 6 had arrived on shore.[5] The wind was blowing with the same fury and it was the general consensus in everybody's mind that the next day in the morning we would no longer see any sign of the Ship. The wind, shall we say, was a little less violent during the night . . . .

A man who did not know how to swim had stayed on board until nightfall and would have perished without the humaneness of a man from a neighboring village. He had attached a bar in a cross configuration at the end of a little Mast, and had sat on it, armed with rope of which some protected him from the violent waves. While 7 or 8 men were advancing the mast by the opposite end, they were able to bring it to the Ship, then this man who was on the piece of wood threw the rope to the one still on board the ship. The latter took the rope and put it around his body and having knotted it he let himself fall in the Sea and thus he was able to be brought to shore. _____[a person's name] who had been Captain of a Ship in the Dutch Navy having learned of this humanitarian act came to get this man. He kept him at his house for several days and gave him a silver Snuffbox on which was engraved the date of this Act. In addition, the Captain took it upon himself to obtain from the Municipality a testimony honoring the man. The rescuer was asked to go to Amsterdam where he was presented a public prize consisting of a silver Box full of money and engraved with the details of his bravery, etc.

Sunday 9 October, the day before the Storm there had come on board the ship two little birds, male and female which I recognized to be Pinson d'Ardennes.[6]

The day after the storm an aquatic, marine bird on the shore was found, named by the English Garnett [Gannet].

The 5 Frimaire, in the year 5 (25 Nov. '96) left Egmond-op.zee [Egmont] and arrived in Amsterdam.[7]

The 6 wrapped and marked Cases and Trunk.

The 7 Dined at Citizen Fousenbarte's house.

The 8 put my Cases on a covered Boat to Bruxelles [Brussels], addressed to the Citizen Endormi. This Boat was to pass through Anvers [Antwerp].

The 9 obtained Passports from the Admiralty for transit of my Collections without having to be searched by the Dutch Inspectors.

The 10 (30 Nov.) Wrote to the Citizens Bosc, Chion, Bussy, the Rev. Nicholas Collin and Gen. Charles Cotesworth Pincknay [Pinckney], by way of New York. Left Amsterdam for Leyde [Leiden] slept in Harlem [Haarlem].[8]

The 11 frimaire (1st Dec. 1796 old style) arrived in Leyde [Leiden] from Harlem [Haarlem] . . . and Amsterdam . . .

Visited Professor of natural history Brugmans to whom I gave several seeds from America. Bought several natural history books.[9]

The 13 frimaire (2 Dec.) left for La Haye [The Hague] and dined the same day at the house of the French Minister of the Dutch Republic.

The 14 (3 Dec.) dined with the Minister.

The 15 left for Roterdam [Rotterdam] 5 hrs from La Haye [The Hague].

The 16 frimaire visited the Brothers Gevers whose Collection of Birds is one of the rarest and best prepared of the ones that I had ever seen.[10]

Visited Doctor Van Noorden, Consul Le Roux la Ville . . .

The 17 left Roterdam [Rotterdam] passed by Dort [Dordrecht] and arrived at Mordick, mouth of many rivers and very dangerous to cross. Slept near Breda, a very fortified town, 9 leagues from Roterdam [Rotterdam].

The 18 arrived in Anvers [Antwerp] 10 leagues from Breda.

The 19 frimaire went for information to the Office of Customs concerning my Cases and Trunks sent from Amsterdam to Bruxelles [Brussels].

The 20 frimaire, The Office was closed. I could not transact any business.

The 21 visited Citizen Bruslé, Comissioner of the Executive Directory and the Citizen Petit-Mongin [also spelled Mangin], Director of Customs. I was very satisfied with the patriotism and National Spirit of Cit. P. Mongin as well as his honesty and strong intellect. I finished my affair concerning the safety and transport of my Cases.

The 22 frimaire left for Bruxelles [Brussels].

The 23 settled with Citizen J. B. Champon's son for the transit of my Cases and Trunks.

The 24 visited the Baron de Reynegom and bought from him Ducks from the Mississippi to replace those that were lost on the Wreck of the 19 Vendemiaire.

The 25 (15 Nov. old style) left Bruxelles [Brussels] for Ghent [Gent], arrived the next morning.

The 26 visited M. Van Aken.

The 27 left for Lille [Lille, France].

The 28 . . .

The 29 . . .

The 30 left Lille.

The 1 Nivose (21 Dec. Wednesday old style) passed by Douay [Douai], Cambray [Cambrai].[11]

The 3 arrived in Paris.

The 4 sent to the National Museum four Ducks (Anas sponsa) from the Mississippi and two Ducks (A. galericulata) from China. Visited Citizens Thouin, Daubenton, Richard, Desfontaines.[12]

The 5 visited Citizens Cels, Tessieu and Andrieux, all three attached to the 4th Division of the Department of the Ministry of Interior Agriculture.[13]

Visited L'Héritier curator of the Botanical directory etc.[14]

The 6 visited Mangourit, Cit. Croix, Minister of external affairs, Colonel Fulton and etc. Assisted at the session of Institut National de France.[15]

Visited Citizens Lamarque [Lamarck], Jussieu etc.[16]

The 7 Wrote to Minister of Interior, to Mangourit, to Chamon [Champon] in Bruxelles. I went to Versailles and slept in Satory.[17]

The 8 Nivose slept and dined at Satory.

The 9 Visited Le Monnier and dined with him.[18]

The 10 Visited L'Héritier and his home with G. Pinckney, dined at Cels.[19]

The 11 Visited Jean Thouin, Mme Gilbert, Mme Le Clerc, Mme Trouve, wife of the editor of Monit. previously Gorelli.[20]

The 12 Visited with General Pinckney, the Garden [Jardin des Plantes] and Natural History collections. Dined with M. Goy and visited M. Barquet.

The 13 looked for lodging.

The 14 visited again L'Héritier, Mr. Dupont and dined with the Directeur de La Reveillière Lepéau.[21]

The 15 public meeting of the Institut National of Sciences and Arts.

The 16 visited Richard, Thouin, meeting of the Institut.

The 17 wrote to Cit. Petit-Mangin, Inspector of Customs in Anvers [Antwerp], and Citizen Champon in Bruxelles [Brussels].

Dined at Remi Claye opposite the pont au change.[22]

The 18 worked on Moving.

The 19 dined at Citizens Redouté, Painters at the Louvre.[23]

The 20 dined at Cels.

The 21 went to the Institut, dissertation of Ventenat on the Phallus [mushroom] of Cayenne.

The 22 Nivose visited the Pantheon.[24]

The 23 bought several pieces of Furniture. Visited Mr. Dubois and Minister Benezech.[25]

The 24 visited Thouin, Delaunay, and Desfontaines. Dined with Mme Barquet.[26]

The 25 had the carpenter do some work. Wrote to Bruxelles [Brussels].

The 26 visited Mangourit: Meeting of the Institut, Dissertation on the Rhinoceros with one and two horns; received a letter of Cit. Petit Mangin; He says that my Collections had not yet arrived in Anvers [Antwerp] on the 22 Nivose. [12 Jan.].

The 27 wrote several letters.

The 28 I went to Thouin; met the Directeur de La Reveillière Lepéau.

The 29 visited Citizen Louvet.

The 30 went to Citizen Cels.

The 1st Pluviose, I went to the National Institut; gave to Cels a letter of the Minister of the Interior to be sent to the Consul in Charleston.

The 2 wrote to Citizen Dupont and sent the letter of the Minister of the Interior.

The 3 I went to the Office of the Minister of the Navy and to Gen. Pinckney, Bernadde [Bernard de] Ste Afrique.[27]

The 4 wrote several letters to: Bosc in duplicate, Capt Baas, Mme. Duverney, Mr. Duverney, Dupont at Charleston, Bussy in New York, Chion, Saulnier.[28]

The 5 Dined at Gen. Pinckney.

The 6 wrote to the Minister of the Navy and sent Papers concerning Spillard, Institut National of Sciences.

The 7 wrote to Himely in Switzerland and to Mme Himely in Charleston.[29]

The 8, 9, 10 worked to put in order the seed collection from Illinois: dined at Cels and gave him a collection of these seeds.[30]

End of Journal

Commentary

Notes

1. Michaux uses two styles of dating. The new style is based on the reformed calendar called the French Republican or Revolutionary calendar, which is calculated from September 22, 1792, the day the Republic was first proclaimed. A week consists of ten days. There are twelve months, each of thirty days. The old style is based on the Gregorian calendar. For most of this part of the *Journal* the new style precedes the old.

On August 13, 1796, André Michaux boarded the ship *Ophir* in the port of Charleston, South Carolina, headed to Amsterdam, Holland, under command of Captain Johnston. The delay in leaving the port is reminiscent of the episode that occurred when Michaux left Charleston to come to Florida. August 13 should be the 26th of the month of Thermidor, instead of the 27th. Likewise, August 16 should be the 29th of Thermidor and September the 15th should be the 29th of the month Fructidor.

2. The Isles of Scilly are an archipelago of more than 150 isles located in the Atlantic Ocean off the southwest tip of England. La Manche is the French equivalent for the English Channel.

3. The *Ophir* had been at sea for fifty-six days when the coast of Holland came into view. For the past eleven years Michaux had experienced diseases, harsh en-

counters with the land, and severe weather. He regarded his own body as nearly indestructible, but now in the midst of a shipwreck, Michaux was reconciled to the fact that his life might soon end.

4. Much of the account of Michaux's rescue written by Hooker (1825), Coulter (1883), and Eifert (1965) is exaggerated and untrue when compared to Michaux's *Journal*. For example, Michaux did not tie himself to a plank nor was his herbarium lashed securely to his body.

5. Only five or six cases had come near shore at this time; all of Michaux's herbarium was saved from the shipwreck in the end. Also, a box containing birds and quadrupeds and all descriptions and memoirs on his plants and animals were saved. However, notebook one that contained the beginning of his *Journal*, a heavy box of birds, a little trunk of personal papers, and all of his clothes and personal effects were lost (Duprat 1957; Savage and Savage 1986).

6. Pinson d'Ardennes are chaffinches of the Ardennes, a series of wooded plateaus and hills in northern France, southeast Belgium, and northern Luxembourg.

7. Michaux remained at Egmont for more than five weeks, during which time he regained his health and restored order to his collection. In a letter dated November 7, 1796, to Charles-Louis L'Héritier de Brutelle, Michaux wrote that he worked for sixteen days from 4 A.M. until 8 P.M. and used seventeen reams of paper. He noted that the colors of his plants had changed, but that no plants were lost (Duprat 1957).

8. Louis Auguste Guillaume Bosc (1759–1828) was at Charleston, but did not take over the care of the Carolina nursery as stated by Savage and Savage (1986). Bosc stayed at the Michaux house and did study and collect many natural history specimens in the Charleston area. He brought seeds from plants in the Michaux garden and gave some of these to Cels. Michaux had been a frequent guest in the Charleston home of General Charles Cotesworth Pinckney (1746–1825). Michaux wrote General Cotesworth Pinckney via New York before the general had left for France; Pinckney was in Paris from December 1796 to February 1797.

9. Sebald Justinus Brugmans (1763–1829) was professor of natural history at Leiden. Because Michaux had lost his books in the shipwreck, he purchased several at Leiden. Beginning with the next day, December 2, and through the statement, "le 30 part. de Lille," the dates for the new style are off by one day. For example, December 2 should be the 12th of Frimaire and not the 13th, and his departure from Lille occurred on December 19 or 29 Frimairie and not 30 Frimaire as given in the *Journal*. Beginning with the first of Nivose (December 21) and continuing to the end of the *Journal*, the dates are correct.

10. This account and others in the *Journal* clearly indicate that Michaux's interest in birds had increased since his earlier days in America.

11. These two towns are just south of Lille, in northern France.

12. The two species of ducks, *Anas sponsa* and *Anas galericulata*, are known today as the wood duck (*Aix sponsa*) and mandarin duck (*Aix galericulata*). Males of both species are highly colored. Michaux was especially fond of the wood duck, or summer duck as he called the species, and a number of living specimens were shipped to France. Michaux visited Louis-Jean Marie Daubenton (1716–1799), French naturalist, anatomist, and collaborator with Comte de Buffon (1707–1788) on his encyclopedic *Histoire Naturelle*, Louis Claude-Marie Richard (1754–1821), and René Louiche Desfontaines (1750–1833). The Thouin mentioned here is either Jean Thouin, brother to André Thouin (1747–1824), or the Thouin household. According to Letouzey (1989), André Thouin had been sent by the government to Italy in May 1796 and he did not return to France until April 1798.

13. Jacques Philippe Martin Cels (1740–1806). Tessieu is most likely Alexandre Henri Tessier (1741–1837), publisher and writer on agricultural subjects in such works as *Dictionnaire d'Agriculture* and *Encyclopédie Méthodique*.

14. Charles-Louis L'Héritier de Brutelle (1746–1800) was a famous French naturalist, prosecutor for the King, and supervisor of waters and forests of Paris. He worked on *Cornus*, *Geranium*, and described *Michauxia*.

15. Michel Ange Bernard Mangourit (1752–1829) was first consul to Charleston under the French Republic. Colonel Robert Fulton (1767–1815) was a famous engineer who worked with Joel Barlow (1754–1812), American ambassador to France, on a large panoramic painting. Fulton became acquainted with famous scientists of the Institut de France, and, according to Savage and Savage (1986), Fulton was seeking financial backing for his invention of the torpedo.

16. Jean-Baptiste de Lamarck (1744–1829) and Antoine-Laurent de Jussieu (1748–1836).

17. Michaux returned to his birthplace of Satory.

18. Louis-Guillaume Le Monnier (1717–1799), Michaux's mentor and principal professor of botany at Jardin du Roi.

19. Here Michaux visited with L'Héritier and with General Pinckney. On the 10th of Frimaire (November 30) Michaux had written a letter to the general by way of New York. General Pinckney was sent to France with John Marshall and Elbridge Gerry when relations with the French Directory had deteriorated to the brink of war in 1796. The general was slated to succeed James Monroe as minister to France, but was rejected by the French government.

20. Jean Thouin, brother to André, became head gardener at the Jardin des Plantes when André was named professor of horticulture in 1793. Madame Gilbert should be Marie-Jeanne Guillebert, oldest sister of André Thouin and head of the Thouin household. Madame Le Clerc was another sister of André Thouin.

Madame Trouvé was André Thouin's natural daughter, by the nursemaid of Buffon's son, who was adopted by the Thouin household at the age of two. She was known as Gorelli before her marriage to Claude Joseph Trouvé, editor-in-chief of the newspaper *Moniteur Universal* (Letouzey 1989).

21. Mr. Dupont was most probably Pierre Samuel Dupont de Nemours (1739–1817), the well-known economist whose son, Victor Dupont, was French Consul in Charleston and later co-founder of an industrial empire in Delaware. Louis Marie Directeur La Revellière Lépeaux (1753–1824) was an amateur botanist. He was influenced by J. J. Rousseau, member of the Revolutionary Consul and Directoire and head of the Constitutional Convention. Lépeaux was a great friend of Thouin and was saved by Louis Auguste Bosc during the French Revolution.

22. Michaux dined with a relative of his deceased wife.

23. Pierre-Joseph Redouté (1759–1840) and his brother H. J. Redouté were painters at the Louvre. The former artist did the illustrations for André Michaux's *Histoire des Chênes de l'Amérique* (1801, Paris) and *Flora Boreali-Americana* (1803, Paris) and François André's *The North American Sylva* (1817–18, Paris). No doubt there was already conversation with the Redoutés regarding the illustrations for Michaux's two works.

24. The Panthéon is a famous monument in Paris that was initially built (1754–1789) as a temple, but after the Revolution the monument became a repository of ashes of famous people.

25. Jean Baptiste Dubois (1753–1808) was a naturalist and economist interested in agriculture. He was a friend of Chrétien Guillaume de Malesherbes (1721–1794) as well as tutor to Malesherbes' grandson.

26. Claude Jean Veau Delaunay (1755–1826) was a student of Louis-Jean Marie Daubenton.

27. Bernardde St^{e} Afrique is incorrect as given in Sargent (1889); Michaux wrote Bernard de St^{e} Afrique in his *Journal*. The identity of this individual is not known; however, it may refer to Jacques Henri Bernadin de Saint-Pierre (1737–1814). He was a writer, naturalist, and had been intendant of the Jardin des Plantes in 1792.

28. Saulnier is Pierre Paul Saunier (ca. 1751–1818), now on his own and in charge of the New Jersey nursery. Usually Michaux used the latter spelling.

29. Jean Jacques Himely, native of Switzerland, was Michaux's friend and patron in Charleston. Himely purchased the Carolina nursery for fifty-three guineas at public auction on March 27, 1792. Himely ceded the property back to Michaux (Savage and Savage 1986). Michaux entrusted the Charleston garden in 1796 to Himely, pending his return to the United States. Himely kept the Charleston garden from going to ruins. Michaux knew that Mr. Himely was in Switzerland and Mme. Himely was in Charleston.

30. At the end of January 1797, Michaux ceased to keep his *Journal*. The final entry dealt with putting in order seeds from Illinois. In that same year, Michaux suffered a very personal loss in the death of Louis-Guillaume Le Monnier, who had been Michaux's benefactor.

Michaux was happy to see his beloved son, François André. It was not long after his arrival in Paris that Michaux learned the devasting reality of the loss of most plants that he had slaved to collect and ship to France during the past eleven years. The royal grounds of Rambouillet were badly scarred from the ravages of the Revolution. Queen Marie Antoinette had sent many of Michaux's plants to enrich the royal gardens of her native Austria (Coats 1969; Savage and Savage 1986). Though disheartening as these events must have been, Michaux's spirit to overcome prevailed. He plunged himself into the writing of his *Histoire des Chênes de l'Amérique* (1801, Paris) and *Flora Boreali-Americana* (1803, Paris); neither would he see in print. The text of the former was finished before he left on the Baudin expedition; only the engravings of the oaks needed to be completed.

With his personal funds exhausted and failing to receive a commission to return to America or to Persia, the future for André Michaux once again looked dismal. After requesting to serve as naturalist on an expedition to examine the coasts of New Holland, now Australia, and New Guinea, in the fall of 1800, he received the offer from the French government and he quickly accepted. He left his son in charge of getting his book on oaks published and completing the publication of the *Flora*.

On October 19, 1800, the botanist, along with a team of scientists, left Le Havre. Because of his desire to continue his research on the rich flora of Île de France, Mauritius today, in the Indian Ocean, and because of internal problems with the stern disciplinarian Captain Nicolas Thomas Baudin (1756–1803), commander of the ill-conducted expedition, the following spring he, most other scientists, and a large number of the crew on board the *Naturaliste*, abandoned the voyage to Australia to remain on the Île de France. Then in the middle of June of 1802, André Michaux sailed to Madagascar. In keeping with his usual behavior, the botanist went right to work to establish a garden and to collect plants for France. Michaux was stricken by a fever. He pulled through the first attack, but the second bout was too much for this hard-working botanist, inveterate explorer, and collector. André Michaux died on October 11, 1803 (Hooker 1825; Dorr 1997). That same year, the bad-tempered Baudin, who scorned scientific work and was indifferent to human suffering, met his own fate from tuberculosis on the Île de France as he was returning to France (Savage and Savage 1986).

André Michaux is believed to be buried at Isatrano on the Ivondro River, west of Mahasoa and south of Tamatave, Madagascar (Dorr 1997). He was fifty-six years old—still in the prime of life.

Thirty years before André Michaux met his fate, Philibert Commerson, royal botanist to King Louis XV, had come to seek the treasures of Madagascar for France. Shortly before this intrepid French botanist died on the island at age forty-six, he wrote the befitting words: "One cannot abstain when in the sight of the rich treasures scattered so freely over this fertile land, from a feeling of pity for those gloomy indoor theorisers who pass their lives in hammering out vain systematics" (Tyler-Whittle 1997).

Appendix 1-1

Plants Documented by John Bartram, William Bartram, and André Michaux while in Florida

John and William Bartram traveled the St. Johns River from November 1765 through February 1766. William's two trips up the river from Spalding's Lower Store to Lake Beresford occurred in May–June and August–September 1774. André Michaux traveled up the St. Johns from April 30 to May 14, 1788. In the accompanying list, species preceded by asterisks were found by the Bartrams in areas, mainly the St. Johns River, traveled later by Michaux. Localities cited were taken from Harper (1942, 1943, 1958). The **M** after a plant name indicates that the species or genus was also found by André Michaux in the same general area explored by the Bartrams. Plant names are those used today and are based on Wunderlin (1998).

Abutilon hulseanum—Near Alachua Savanna (Paynes Prairie of today), Alachua Co.

**Acer negundo*—On St. Johns River above Lake Dexter; at Crows Bluff, Lake Co.; near Alachua Savanna (Paynes Prairie of today), Alachua Co.

**Acer rubrum* (M)—Salt Springs Run; St. Johns River near Rollestown, Putnam Co.; on Johns River above Lake Dexter; near Lake Beresford, Volusia Co.; near Lochloosa Lake, S of Hawthorn, Alachua Co. near Alachua Savanna (Paynes Prairie of today), Alachua Co.

Aesculus pavia—Near scrub at Salt Springs, Marion Co.; near Alachua Savanna (Paynes Prairie of today), Alachua Co.

Agalinis sp.—Near Suwannee River; near Long Pond, S of Chiefland, Levy Co.

Agarista populifolia (M)—Near Deep Creek, N of Kenwood, Putnam Co.; near Palatka, Putnam Co.; about St. Marys River, GA-FL.

**Agave decipiens*—On Mosquito River (Mosquito Lagoon of today).

Agave sp.—Near Long Pond, S of Chiefland, Levy Co.

Aletris sp.—Near Long Pond, S of Chiefland, Levy Co.

**Amaranthus* sp. (M)—Along St. Johns River above Lake Dexter.

Amianthium muscaetoxicum—Near Long Pond, S of Chiefland, Levy Co.

**Ampelopsis arborea*—On St. Johns River above Lake Dexter.

Andromeda sp.—Near Palatka, Putnam Co.; near Alachua Savanna (Paynes Prairie of today), Alachua Co.; near Orange Lake, Alachua Co.

**Apios americana*—Near Lochloosa Lake, S of Hawthorn, Alachua Co.; on St. Johns River above Lake Dexter.

Arisaema sp.—Between Alachua Savanna and Suwannee River.

Aristida beyrichiana (?)—Between Alachua Savanna and Suwannee River.

Aristolochia sp.—Near Alachua Savanna (Paynes Prairie of today), Alachua Co.

**Arundinaria gigantea* (M)—On Lake Dexter, Volusia Co.; Alachua Savanna (Paynes Prairie of today), Alachua Co.

Asarum arifolium—Manatee Springs, Levy Co.

Asclepias humistrata—Near Suwannee River.

Asimina incana—Near Rodman, Putnam Co.

**Asimina obovata* (M)—Near Rollestown, Putnam Co. Harper's (1958) *Asimina incana* of large, white, fragrant flowers probably should be *A. obovata*.

Asimina pygmaea—Near Cuscowilla (Seminole Indian town near present Micanopy), Alachua Co.; near Orange Lake, Alachua Co.; near Rodman, Putnam Co.; W of Spalding's Lower Store, Putnam Co.

Asimina sp.—Near Alachua Savanna (Paynes Prairie of today), Alachua Co.; near Palatka, Putnam Co.

**Aster carolinianus*—Along St. Johns River above Lake Dexter.

**Aster* sp. Near Suwannee River; near Lake Beresford, Volusia Co.; between Alachua Savanna and Suwannee River.

**Avicennia germinans*—Mosquito River (Mosquito Lagoon of today).

**Baccharis halimifolia*—On St. Johns River above Lake Dexter.

Baptisia sp. (M)—Near Lochloosa Lake, S of Hawthorn, Alachua Co.; between Orange and Tuscawilla Lakes, Alachua Co. near Suwannee River; near Long Pond, S of Chiefland, Levy Co.

Bejaria racemosa (M)—W of Spalding's Lower Store, Putnam Co. near Palatka, Putnam Co.

Berchemia scandens (M)—On lower St. Johns River; about Alachua Savanna (Paynes Prairie of today), Alachua Co.

**Bidens* sp. (probably *B. mitis*)—On floating islets of the St. Johns River.

**Bignonia capreolata*—On lower St. Johns River; St. Johns River above Lake Dexter.

**Brunnichia ovata*—On St. Johns River above Lake Dexter.

Calamintha coccinea—Near Suwannee River; Pensacola, Escambia Co.

**Callicarpa americana*—Near scrub at Salt Springs, Marion Co.; on St. Johns River above Lake George; between Orange and Tuscawilla Lakes, Alachua Co.; near Alachua Savanna (Paynes Prairie of today), Alachua Co.; between Alachua Savanna and Suwannee River; in Levy Co.; at Manatee Springs, Levy Co.

**Calydorea caelestina* (M)—Near the shore of Lake Dexter, Volusia Co.; W of Kanapaha Prairie, Alachua Co.

**Campanula floridana*—At Mount Royal, S of Fruitland, Putnam Co.

Campsis radicans—On lower St. Johns River N of Rollestown, Putnam Co.

**Canna flaccida* (M)—At Lake Dexter, Volusia Co.

Capsicum frutescens—Nassau Sound, Duval Co.

**Carica papaya* (M)—On St. Johns River at Rollestown and above Lake Dexter.

Carpinus caroliniana—Near Alachua Savanna (Paynes Prairie of today), Alachua Co.; between Alachua Savanna and Suwannee River.

**Carya cordiformis*—On St. Johns River above Lake Dexter; between Orange and Tuscawilla Lakes, Alachua Co.; near Alachua Savanna (Paynes Prairie of today), Alachua Co.

**Carya* sp.—Near Alachua Savanna (Paynes Prairie of today), Alachua Co.; near Cuscowilla (Seminole Indian town near present Micanopy), Alachua Co.; at Manatee Springs, Levy Co.; on St. Johns River at or near Crows Bluff, Lake Co.

**Castanea* sp.—Along St. Johns River near Rollestown.

**Catalpa bignonioides*—Along St. Johns River above Lake Dexter.

**Celtis laevigata*—On St. Johns River at Mosquito Grove, Lake Co., and at Lamb's Bluff (near Lake Beresford); near Alachua Savanna (Paynes Prairie of today), Alachua Co.

**Cephalanthus occidentalis*—On St. Johns River above Lake Dexter and near Lake Beresford, Volusia Co.

Ceratiola ericoides—W of Spalding's Lower Store, Putnam Co.; near Orange Lake, Alachua Co.; near Long Pond, S of Chiefland, Levy Co.; near Suwannee River.

Cercis canadensis—Near Alachua Savanna (Paynes Prairie of today), Alachua Co.

**Chaptalia tomentosa*—At Lake Dexter.

Chionanthus virginicus—Near Deep Creek, N of Kenwood, Putnam Co.; about St. Marys River, GA-FL.

Cirsium horridulum—Near Deep Creek, N of Kenwood, Putnam Co.

Citrullus lanatus—(cultivated)—At Palatka, Putnam Co.

**Citrus aurantium* (M)—Along Mosquito Lagoon; at Nassau Sound, Duval Co.; at Palatka, Putnam Co.; on St. Johns River above Jacksonville to Blue Spring, Volusia Co.; on Drayton Island, Putnam Co.; in Alachua Co.; at Little Orange Lake, Alachua-Putnam Cos.; about Alachua Savanna (Paynes Prairie of today), Alachua Co.

**Citrus sinensis* (M)—Along Mosquito Lagoon; at Nassau Sound, Duval Co.; at Palatka, Putnam Co.; on St. Johns River above Jacksonville to Blue Spring, Volusia Co.; on Drayton Island, Putnam Co.; in Alachua Co.; at Little Orange Lake, Alachua-Putnam Cos.; about Alachua Savanna (Paynes Prairie of today), Alachua Co.

Clethra alnifolia—Near Deep Creek, N of Kenwood, Putnam Co.; about St. Marys River, GA-FL.

**Clitoria mariana*—St. Johns River at Lake Dexter, Volusia Co. W of Deep Creek, N of Kenwood, Putnam Co.; near Long Pond, S of Chiefland, Levy Co.; near Lochloosa Lake, S of Hawthorn, Alachua Co.

Cnidoscolus stimulosus—Near Long Pond, S of Chiefland, Levy Co.

Collinsonia sp.—Near Alachua Savanna (Paynes Prairie of today), Alachua Co.

Cornus florida—Near Alachua Savanna (Paynes Prairie of today), Alachua Co.

**Cornus foemina*—Near Alachua Savanna (Paynes Prairie of today), Alachua Co.; on St. Johns River above Lake Dexter.

**Crinum americanum* (M)—At Lake George.

Cucurbita moschata (cultivated)—Near Talahasochte (Seminole Indian town on Suwannee River), Levy Co.; at Palatka, Putnam Co.

**Cucurbita okeechobeensis* (M)—On St. Johns River above Lake Dexter.

Cucurbita pepo (cultivated)—Near Alachua Savanna (Paynes Prairie of today), Alachua Co.

Cucurbita verrucosa (cultivated)—Near Alachua Savanna (Paynes Prairie of today), Alachua Co.; near Palatka, Putnam Co.

Cyrilla racemiflora—Near Deep Creek, N of Kenwood, Putnam Co.; about St. Marys River, GA-FL.

Dalea sp.—Near Suwannee River.
Desmodium sp. or *Lespedeza* sp.—Near Suwannee River.
**Diospyros virginiana*—On St. Johns River above Lake Dexter.

Elephantopus sp. (?)—Near Suwannee River.
**Erythrina herbacea* (M)—On Mosquito River (Mosquito Lagoon today); near Long Pond, S of Chiefland, Levy Co.; scrub near Salt Springs, Marion Co.; near Silver Glen Spring, Marion Co.; W of Kanapaha Prairie, Alachua Co.; near Halfway Pond (Cowpen Lake of today), Putnam Co.
Euonymus americanus—Near Alachua Savanna (Paynes Prairie of today), Alachua Co.
Eupatorium sp.—Near Alachua Savanna (Paynes Prairie of today), Alachua Co.; between Alachua Savanna and Suwannee River; Manatee Springs, Levy Co.
Euphorbia sp.—Near Suwannee River.

Fagus grandifolia—Near Alachua Savanna (Paynes Prairie of today), Alachua Co.; W of the Suwannee River; between Orange and Tuscawilla Lakes, Alachua Co.; on lower St. Johns River; near Lochloosa Lake, S of Hawthorn, Alachua Co.
**Fraxinus americana*—On St. Johns River at Rollestown, Putnam Co.; on St. Johns River above Lake Dexter; near Alachua Savanna (Paynes Prairie of today), Alachua Co.
**Fraxinus caroliniana* (M)—About St. Marys River, GA-FL; on St. Johns River above Lake Dexter.
**Fraxinus* sp.—St. Johns River, near Rollestown, Putnam Co.; on Salt Springs Run, Marion Co. W of Suwannee River.

Garberia heterophylla—Scrub near Salt Springs, Marion Co.; on Mosquito River (Mosquito Lagoon today).
**Gleditsia aquatica* (*Gleditsia tricanthos*) (M)—On St. Johns River above Picolata; on St. Johns River above Lake Dexter; at or near Crows Bluff, Lake Co.; near Lake Beresford, Volusia Co.
**Gnaphalium obtusifolium*—At Lake Dexter.
**Gordonia lasianthus* (M)—About St. Marys River, GA-FL; on St. Johns River near Rollestown, Putnam Co.; at Salt Springs, Marion Co.; about lakes in Putnam Co.; near Deep Creek, N of Kenwood, Putnam Co.
**Gossypium barbadense* (cultivated)—At Lake George.

Gossypium hirsutum (cultivated)—On St. Johns River; at Mount Royal, S of Fruitland, Putnam Co.

Halesia carolina (M)—Near scrub at Salt Springs, Marion Co.; near Alachua Savanna (Paynes Prairie of today), Alachua Co.

Halesia diptera—Near scrub at Salt Springs, Marion Co.; near Alachua Savanna (Paynes Prairie of today), Alachua Co.

Halesia sp.—Near Alachua Savanna (Paynes Prairie of today), Alachua Co.; between Alachua Savanna and Suwannee River; at Manatee Springs, Levy Co.; near Suwannee River.

Helenium sp.—Near Suwannee River; near Alachua Savanna (Paynes Prairie of today), Alachua Co.; at Manatee Springs, Levy Co.

Helianthemum sp.—W of Kanapaha Prairie, Alachua Co.; near Suwannee River, FL; near Long Pond, S of Chiefland, Levy Co.

**Helianthus* sp.—Near Suwannee River; near Lochloosa Lake, S of Hawthorn, Alachua Co.; W of Kanapaha Prairie, Alachua Co.

**Hibiscus coccineus* (M)—On St. Johns River above Lake Dexter; on Drayton Island.

**Hibiscus moscheutos*—On St. Johns River above Lake Dexter; on Drayton Island.

**Hydrocotyle* sp. or *Centella* sp. (M)—On Drayton Island.

**Hymenocallis* sp. (M)—On Drayton Island; along St. Johns River above Lake Dexter.

Hypoxis sp.—Near Long Pond, S of Chiefland, Levy Co.

**Ilex cassine* (M)—About St. Marys River, GA-FL; near Suwannee River; near Silver Glen Spring, Marion Co.

**Ilex glabra* (M)—Near Alachua Savanna (Paynes Prairie of today), Alachua Co.; near Silver Glen Spring, Marion Co.

Ilex opaca (M)—Near scrub at Salt Springs, Marion Co.; lower St. Johns River, Putnam Co.; near Suwannee River.

**Ilex vomitoria* (M)—Near Salt Springs; W of Spalding's Lower Store; near Alachua Savanna (Paynes Prairie of today), Alachua Co.

**Illicium parviflorum* (M)—At Salt Springs, Marion Co.; at Silver Glen Spring, Marion Co.

Indigofera caroliniana—In Levy Co.; between Orange and Tuscawilla Lakes, Alachua Co.

**Indigofera suffruticosa* (introduced and cultivated).

**Indigofera tinctoria* (introduced and cultivated).

Ipomoea alba (M)—On Lake George and Drayton Island; on St. Johns River above Lake Dexter.

Ipomoea batatas—On lower St. Johns River; at Palatka, Putnam Co.

**Ipomoea* sp.—W of Deep Creek, N of Kenwood, Putnam Co.; near Lochloosa Lake, S of Hawthorn, Alachua Co.; Lake George; on St. Johns River near Lake Beresford, Volusia Co.

Itea virginica—About St. Marys River, GA-FL.

Juncus sp. (?) About Alachua Savanna (Paynes Prairie of today), Alachua Co.

**Juniperus virginiana* (M)—At Hontoon Island on St. Johns River.

**Kalmia hirsuta*—Near the seacoast of Florida; near Orange Lake, Alachua Co.; on St. Johns River at Rollestown, Putnam Co.; W of Spalding's Lower Store, near Suwannee River.

**Kosteletzkya virginica*—On Drayton Island, Putnam Co.

Krameria lanceolata—Near Alachua Savanna (Paynes Prairie of today), Alachua Co.; near Suwannee River.

**Lagenaria siceraria* (cultivated)—on St. Johns River.

**Lantana camara*—At Mount Royal, S of Fruitland, Putnam Co. on Drayton Island.

Lechea sp.—Near Suwannee River.

Lespedeza sp. (?)—Near Suwannee River.

Leucothoe axillaris—Near Palatka, Putnam Co.

Lilium catesbaei—Near Long Pond, S of Chiefland, Levy Co.

**Liquidambar styraciflua*—On St. Johns River at Lamb's Bluff (near Lake Beresford); near Alachua Savanna (Paynes Prairie of today), Alachua Co.

Liriodendron tulipifera—Near Tuscawilla Lake, Alachua Co.

**Lobelia cardinalis*—On Drayton Island, Putnam Co.

Lobelia sp.—Near Alachua Savanna (Paynes Prairie of today), Alachua Co.; near Suwannee River.

Lupinus perennis—Near Suwannee River.

Lupinus sp.—About St. Marys River, GA-FL.

Lycium carolinianum—Amelia Island, Nassau Co.

Lyonia ferruginea (M)—About St. Marys River, GA-FL; W of Spalding's Lower Store, Putnam Co.; Manatee Springs, Levy Co.

Lyonia lucida—Near Suwannee River; in Putnam Co.; about St. Marys River, GA-FL.

*_Magnolia grandiflora_ (M)—At Lake George; St. Johns River above Lake Dexter; at Hawkinsville; near Tuscawilla Lake, Alachua Co.; at Salt Springs, Marion Co.; at Silver Glen Spring, Marion Co.; St Johns River; below Palatka, Putnam Co.; above Lake George; near Hontoon Island; near Alachua Savanna (Paynes Prairie of today), Alachua Co.; Suwannee River; Manatee Springs, Levy Co.; near Cuscowilla (Seminole Indian town near present Micanopy), Alachua Co.

*_Magnolia virginiana_ (M)—Near scrub at Salt Springs, Marion Co.; near Deep Creek, N of Kenwood, Putnam Co.; about St. Marys River, GA-FL; between Alachua Savanna and Suwannee River; along Camp Branch, Putnam Co.

*_Merremia dissecta_ (M)—On Drayton Island, Putnam Co.

*_Mikania scandens_—On St. Johns River above Lake Dexter.

Mitchella repens—Near Alachua Savanna (Paynes Prairie of today), Alachua Co.

*_Morus rubra_—On St. Johns River at Rollestown, Putnam Co.; above Lake Dexter, Volusia Co.; at or near Crows Bluff, Lake Co.; near Salt Springs, Marion Co.; between Orange and Tuscawilla Lakes, Alachua Co.; near Alachua Savanna (Paynes Prairie of today), Alachua Co.; in Levy Co.; near Suwannee River; between Alachua Savanna and Suwannee River; near Cuscowilla (Seminole Indian town near present Micanopy), Alachua Co.

*_Myrica cerifera_ (M)—About St. Marys River, GA-FL; on St. Johns River above Lake George; near Orange Lake, Alachua Co.; near Nassau Sound, Duval Co.; near Alachua Savanna (Paynes Prairie of today), Alachua Co.

*_Nelumbo lutea_—At Lake George; near Lake Dexter; in Suwannee River; in Chacala Pond, Alachua Co.; in Suwannee River; in East Florida.

Nuphar lutea spp. _advena_—Near Alachua Savanna (Paynes Prairie of today), Alachua Co.

Nymphaea odorata—Near Alachua Savanna (Paynes Prairie of today), Alachua Co.

Nyssa aquatica (M)—About St. Marys River, GA-FL; on St. Johns River near Rollestown; on St. Johns River above Lake Dexter; at Salt Springs, Marion Co.; near Deep Creek, N of Kenwood, Putnam Co.

Nyssa ogeche—Near Suwannee River; about St. Marys River, GA-FL

*_Nyssa_ sp.—On Salt Springs Run; on St. Johns River near Rollestown.

*_Nyssa sylvatica_—About St. Marys River, GA-FL; on St. Johns River above Lake Dexter; at Salt Spring; near Deep Creek, N of Kenwood, Putnam Co.

Nyssa sylvatica var. *biflora*—Between Alachua Savanna (Paynes Prairie of today), Alachua Co.; along Suwannee River.

Onosmodium virginianum—Near Long Pond, S of Chiefland, Levy Co.

**Opuntia* sp.—Near Suwannee River; near Long Pond, S of Chiefland, Levy Co.; W of Kanapaha Prairie, Alachua Co.; near Halfway Pond (Cowpen Lake of today), Putnam Co.

Orontium aquaticum—Near Alachua Savanna (Paynes Prairie of today), Alachua Co.

Oryza sativa (cultivated)—Near Rollestown, Putnam Co.; Lake Beresford, Volusia Co.

**Osmanthus americanus* (M)—About St. Marys River, Ga.–Fla.; on St. Johns River below Palatka, Putnam Co.; on St. Johns River above Lake Dexter; at Lake George; near scrub at Salt Springs; W of Spalding's Lower Store; at Manatee Springs, Levy Co.; near Alachua Savanna (Paynes Prairie of today), Alachua Co.

Osmunda sp.—Near Deep Creek, N of Kenwood, Putnam Co.

Oxydendrum arboreum (M)—Near Alachua Savanna (Paynes Prairie of today), Alachua Co.

**Panicum hemitomom* (?)—Near Deep Creek, N of Kenwood, Putnam Co.; on Alachua Savanna (Paynes Prairie of today), Alachua Co.; in outlets of Salt Springs, Marion Co.

**Parthenocissus quinquefolia*—Near Long Pond, S of Chiefland, Levy Co.; on St. Johns River above Lake Dexter; near Alachua Savanna (Paynes Prairie of today), Alachua Co.

**Passiflora incarnata*—At Lake Dexter, Volusia Co.

**Pavonia spinifex*—On St. Johns River above Lake Dexter; on Drayton Island; near Silver Glen Spring.

Pedicularis canadensis—Near Long Pond, S of Chiefland, Levy Co.

**Persea borbonia* var. *borbonia* or *P. palustris* (M)—At St. Francis, Mosquito Grove, Lamb's Bluff, and Crows Bluff on St. Johns River; New Smyrna, Volusia Co.; Salt Spring Run; about Alachua Savanna (Paynes Prairie of today), Alachua Co.; Manatee Springs, Levy Co.; near Suwannee River; between Alachua Savanna and Suwannee River; Nassau Sound, Duval Co.; along Camp Branch, Putnam Co.; in swamps on St. Johns River.

Persea borbonia var. *humilis*—Scrub near Salt Springs, Marion Co.

Phaseolus sp. (M)—Near Alachua Savanna (Paynes Prairie of today), Alachua Co.; near Palatka, Putnam Co.; Suwannee River; between Alachua Savanna and Suwannee River.

**Phlox* sp. (M)—At Lake Dexter, Volusia Co.; W of Deep Creek, N of Kenwood, Putnam Co.; W of Kanapaha Prairie, Alachua Co.

**Phragmites australis*—At Lake Beresford, Volusia Co.

Pinckneya bracteata—In East Florida.

**Pinus glabra*—Silver Glen to Lisk Point, Marion Co.

**Pinus palustris*—At Cowpen Lake, Putnam Co.; near Tuscawilla Lake, Alachua Co.; at Alachua Savanna (Paynes Prairie of today), Alachua Co.; on St. Johns River at Rollestown; W of Deep Creek, N of Kenwood, Putnam Co.; in Escambia Co.; near Salt Springs, Marion Co.; in Levy Co.; W of Spalding's Lower Store; near Rodman, Putnam Co.; W of Kanapaha Prairie, Alachua Co.; between Alachua Savanna and Suwannee River; near Long Pond, S of Chiefland, Levy Co.

Pinus serotina—Near Deep Creek, N of Kenwood, Putnam Co.; near Alachua Savanna (Paynes Prairie of today), Alachua Co.

Pinus sp.—Between Alachua Savanna and Suwannee River.

Pinus taeda—Near Suwannee River.

**Pistia stratiotes* (M)—At Lake Dexter; on St. Johns River above Lake Dexter and below Lake Beresford, Volusia Co.; on Lake George; on Salt Springs Run; on Suwannee River.

Poinsettia heterophylla (M)—On Amelia Island, Nassau Co.

Polygala sp.—Near Suwannee River.

**Polygonum* sp.—On St. Johns River above Lake Dexter.

Polymnia uvedalia—Near Alachua Savanna (Paynes Prairie of today), Alachua Co.; near Lochloosa Lake, S of Hawthorn, Alachua Co.; between Orange and Tuscawilla Lakes, Alachua Co.

Prenanthes sp.—Near Suwannee River.

Prunus caroliniana—About Alachua Savanna (Paynes Prairie of today), Alachua Co.; lower St. Johns River, Putnam Co.; about St. Marys River, Ga.-Fla.

Prunus persica (cultivated)—On St. Marys River, Ga.-Fla.; at Palatka, Putnam Co.

Prunus serotina—About Alachua Savanna (Paynes Prairie of today), Alachua Co.

**Ptelea trifoliata*—On St. Johns River at Rollestown, Putnam Co.; at Lake George; near Salt Springs; between Orange and Tuscawilla Lakes, Alachua Co.; between Alachua Savanna and Suwannee River; at Manatee Springs, Levy Co., near Suwannee River.

Quercus chapmanii—In Putnam Co.; between Alachua Savanna and Suwannee River; Silver Glen Spring, Marion Co.; scrub near Salt Springs, Marion Co.; W of Spalding's Lower Store.

Quercus falcata—E of Alachua Savanna (Paynes Prairie of today), Alachua Co.; W of Deep Creek, N of Kenwood, Putnam Co.

Quercus geminata—Silver Glen Spring, Marion Co.; scrub near Salt Springs, Marion Co.; in Putnam Co.; W of Spalding's Lower Store.

Quercus laevis—Between Alachua Savanna and Suwannee River.

**Quercus laurifolia*—In Liberty Co., about St. Marys River, GA-FL; on St. Johns River below Palatka; on St. Johns River above Lake Dexter; at Salt Springs, Marion Co.; at Alachua Sink.

Quercus marilandica—W of Deep Creek, N of Kenwood, Putnam Co.

**Quercus michauxii*—Near Salt Springs, Marion Co.

Quercus myrtifolia—Between Alachua Savanna and Suwannee River; scrub near Salt Springs, Marion Co.; near Suwannee River; W of Spalding's Lower Store.

**Quercus nigra* (M)—Near Cuscowilla (Seminole Indian town near present Micanopy), Alachua Co.; W of Deep Creek, N of Kenwood, Putnam Co.; about St. Marys River, Ga.–Fla.; in swamps on the St. Johns River; between Alachua Savanna and Suwannee River; at Manatee Springs, Levy Co.

**Quercus phellos* (M)—About St. Marys River, Ga.–Fla.; on St. Johns River above Lake Dexter.

Quercus pumila—Near Alachua Savanna (Paynes Prairie of today), Alachua Co.; scrub near Salt Springs, Marion Co.; between Alachua Savanna and Suwannee River.

**Quercus* sp.—At Palatka, Putnam Co.; on St. Johns near Rollestown, Putnam Co.; on St. Johns at St. Francis, Hawkinsville, near Hontoon Island; at Crows Bluff; Blue Spring, Volusia Co.; near Tuscawilla Lake, near Long Pond, S of Chiefland, Levy Co.; near Suwannee River; between Alachua Savanna and Suwannee River; near Alachua Savanna (Paynes Prairie of today), Alachua Co.; Lake Beresford, Volusia Co., near Kanapaha Prairie, Alachua Co.; Manatee Springs, Levy Co.

Quercus velutina (?)—Near Suwannee River.

**Quercus virginiana* (M)—On lower St. Johns River; on St. Johns River at Rollestown, Putnam Co.; about St. Marys River, Ga.–Fla.; on Amelia Island, Nassau Co.; at Nassau Sound, Duval Co.; at Mount Royal, Putnam Co.; on Lake Dexter; on St. Johns River; at Mosquito Grove, Lake Co.; St. Francis, Lake Co.; Bluffton, Lake Beresford, Volusia Co.; Hontoon

Island; at New Smyrna, Volusia Co.; near Salt Springs, Marion Co.; in Putnam County; near Tuscawilla Lake; near Alachua Savanna (Paynes Prairie of today), Alachua Co.; at Kanapaha Sink, Alachua Co.; in Levy County; at Silver Glen Spring, Marion Co.; at Spalding's Lower Store; between Alachua Savanna and Suwannee River; near Cuscowilla (Seminole Indian town near present Micanopy), Alachua Co.

Rhamnus carolinana—About St. Marys River, Ga.–Fla.; at Lake George; near Salt Springs; W of Spalding's Lower Store; between Orange and Tuscawilla Lakes, Alachua Co.; near Suwannee River.
Rhexia sp.—Near Lochloosa Lake, S of Hawthorn, Alachua Co.
**Ricinus communis*—Along St. Johns River above Lake Dexter.
**Roystonea regia*—On St. Johns River between Astor and Lake Dexter.
Rudbeckia fulgida—In Florida.
Rudbeckia sp.—Near Lochloosa Lake, S of Hawthorn, Alachua Co.; between Orange and Tuscawilla Lakes, Alachua Co.
Ruellia sp.—W of Deep Creek, N of Kenwood, Putnam Co.; near Lochloosa Lake, S of Hawthorn, Alachua Co.

**Sabal minor*—On St. Johns River near Rollestown, Putnam Co.
**Sabal palmetto* (M)—on Amelia Island, Nassau Co.; at Nassau Sound, Duval Co.; on St. Johns River above Jacksonville, Duval Co., to Blue Spring, Volusia Co.; in Alachua County near Tuscawilla Lake; about Alachua Savanna (Paynes Prairie of today), Alachua Co.; near Long Pond, S of Chiefland, Levy Co.; Manatee Springs, Levy Co.; near Deep Creek, N of Kenwood, Putnam Co.; at Silver Glen Spring, Marion Co.; between Alachua Savanna and Suwannee River.
Saccharum officinarum (cultivated) (M)—At Lake Beresford, Volusia Co.
**Sagittaria* sp.—On St. Johns River.
**Salix* sp.—Near Alachua Savanna (Paynes Prairie of today), Alachua Co.; St. Johns River above Lake Dexter.
Salvia sp.—Near Lochloosa Lake, S of Hawthorn, Alachua Co.; between Orange and Tuscawilla Lakes, Alachua Co.
**Sambucus canadensis*—In swamps on the St. Johns River; on St. Johns River near Rollestown, Putnam Co., and Lake Beresford, Volusia Co.; between Orange and Tuscawilla Lakes, Alachua Co.
**Sapindus saponaria* (M)—On St. Johns River below Palatka; at Rollestown, Putnam Co.; Alachua Sink, Alachua Co.
Sarracenia leucophylla—Near Pensacola, Escambia Co.

Schrankia sp.—Near Suwannee River, Levy Co.; about St. Marys River, GA-FL.

**Scirpus* sp.—Along Suwannee and St. Johns Rivers.

**Senecio glabellus*—Along St. Johns River above Lake Dexter, Volusia Co.

**Serenoa repens* (M)—Near Alachua Savanna (Paynes Prairie of today), Alachua Co.; near Orange Lake, Alachua Co.; W of Spalding's Lower Store; W of Deep Creek, N of Kenwood, Putnam Co.

**Sideroxylon* sp.—At Lake George; near Salt Springs, Marion Co.; near Alachua Savanna (Paynes Prairie of today), Alachua Co.; in Levy Co.

Silphium sp.—Near Suwannee River; near Lochloosa Lake, S of Hawthorn, Alachua Co.; between Orange and Tuscawilla Lakes, Alachua Co.

Sisymbrium sp.—Near Suwannee River.

Smilax bona-nox—Near Talahasochte (Seminole Indian town on Suwannee River), Levy Co.

Smilax pumila—Near Suwannee River.

Smilax sp.—Near Suwannee River.

Solidago sp.—Near Suwannee River; at Manatee Springs, Levy Co.; on St. Johns River.

Staphylea trifolia—Near Alachua Savanna (Paynes Prairie of today), Alachua Co.

Stewartia malachodendron—Near Alachua Savanna (Paynes Prairie of today), Alachua Co.

**Symplocos tinctoria*—About St. Mary's River, Ga.-Fla.; on St. Johns River above Lake Dexter; near Salt Springs, Marion Co.; near Alachua Savanna (Paynes Prairie of today), Alachua Co.

**Taxodium* sp. (M)—On lower St. Johns River; on St. Johns River above Lake Dexter; about St. Mary's River, Ga.-Fla.; in coastal swamps.

Tephrosia sp.—Near Suwannee River, FL.

Tetragonotheca helianthoides—Near Alachua Savanna (Paynes Prairie of today), Alachua Co.

**Tilia americana*—Between Alachua Savanna and Suwannee River; on St. Johns River at Rollestown; near Tuscawilla Lake, Alachua Co.; near Alachua Savanna (Paynes Prairie of today), Alachua Co.; in Levy Co.; near Cuscowilla (Seminole Indian town near present Micanopy), Alachua Co.

**Tillandsia usneoides*—On lower St. Johns River; near East Tocoi, St. Johns Co.; at Lake Dexter.

Tradescantia sp.—Near Long Pond, S of Chiefland, Levy Co.

**Typha* sp.—At Salt Springs, Marion Co.

**Ulmus alata*—On St. Johns River above Lake Dexter.

**Ulmus americana*—On St. Johns River above Lake Dexter; near Alachua Savanna (Paynes Prairie of today), Alachua Co.

**Ulmus* sp.—Near Cuscowilla (Seminole Indian town near present Micanopy), Alachua Co.; about Alachua Savanna (Paynes Prairie of today), Alachua Co.; in swamps on St. Johns River; St. Johns River near Lake Beresford and Crows Bluff; on St. Johns River below Palatka; at Rollestown; between Alachua Savanna and Suwannee River.

Urtica sp.—Near Alachua Savanna (Paynes Prairie of today), Alachua Co.; at Manatee Springs, Levy Co.; between Alachua Savanna and Suwannee River; near Lochloosa Lake, S of Hawthorn, Alachua Co.

Uvularia perfoliata—At Manatee Springs, Levy Co.

**Vaccinium* sp., or *Gaylussacia* sp.—W of Spalding's Lower Store, Putnam Co.; near Orange Lake, Alachua Co.

**Verbena* sp.—Near Lochloosa Lake, S of Hawthorn, Alachua Co.; at Lake Dexter; W of Deep Creek, N of Kenwood, Putnam Co.; near Long Pond, S of Chiefland, Levy Co.

Vernonia sp. (?)—Near Suwannee River.

Vicia sp.—Between Alachua Savanna and Suwannee River; near Lochloosa Lake, S of Hawthorn, Alachua Co.

Vigna luteola—Near Alachua Savanna (Paynes Prairie of today), Alachua Co.

**Viola* sp.—W of Deep Creek, N of Kenwood, Putnam Co.; near Lochloosa Lake, S of Hawthorn, Alachua Co.; at Lake Dexter.

**Vitis aestivalis*—On St. Johns River above Lake Dexter; near Alachua Savanna (Paynes Prairie of today), Alachua Co.

**Vitis rotundifolia*—On St. Johns River above Lake Dexter.

Vitis sp.—Lower St. Johns River; W of Alachua Savanna (Paynes Prairie of today), Alachua Co.; W of Kanapaha Prairie.

**Wisteria frutescens*—On St. Johns River above Lake Dexter.

**Ximenia americana*—On St. Johns River below Lake Dexter; on St. Johns River at Rollestown.

**Yucca aloifolia*—Near Nassau Sound, Duval Co.; at Lake George.

**Zamia pumila* (M)—Near Long Pond, S of Chiefland, Levy Co.; near Salt Springs.

**Zanthoxylum clava-herculis*—On lower St. Johns River; at Lake George; on St. Johns River at Hontoon Island; at Lake Dexter; near Salt Springs; between Orange and Tuscawilla Lakes, Alachua Co.; near Alachua Savanna (Paynes Prairie of today), Alachua Co.; in Levy Co.

Zephyranthes atamasco—W of Deep Creek, N of Kenwood, Putnam Co.; in Levy Co.; about St. Marys River, Ga.–Fla..

Zigadenus densus—Near Long Pond, S of Chiefland, Levy Co.

Florida Plants Found by André Michaux in the Spring of 1788 and Not Documented by the Bartrams During Their Florida Explorations

This list is based on Michaux's *Journal*, the IDC, the *Flora*, and lists of seeds and plants shipped to France. Current names are given.

Acrostichum danaeifolium
Amaranthus australis
Amorpha fruticosa
Amyris elemifera
Annona glabra
Arnoglossum sp.
Asclepias sp. (shrub)
Berlandiera pumila
Blechnum serrulatum
Blutaparon vermiculare
Caesalpinia bonduc
Canavalia ensiformis
Carya aquatica
Ceanothus microphyllus
Chamaesyce maculata
Chiococca alba
Cochlosperma religiosum (cultivated)
Coelorachis cylindrica
Conocarpus erectus
Croton argyranthemus
Cynanchum sp.
Dalea carnea
Eriocaulon sp.
Eriochloa michauxii
Ficus aurea
Fimbristylis spadicea

Forestiera segregata
Fuirena scirpoidea
Gelsemium sempervirens
Hedyotis nigricans
Hedyotis procumbens
Hippomane sp.
Hymenocallis latifolia
Hypericum myrtifolium
Hypericum tetrapetalum
Ipomoea lacunosa
Ipomopsis rubra
Juncus roemerianus
Laguncularia racemosa
Leucothoe racemosa
Lupinus diffusus
Lyonia fruticosa
Lyonia ligustrina
Lyonia mariana
Melica mutica
Mimosa sp.
Modiola caroliniana
Ocotea coriacea
Paspalum praecox
Pecluma plumula
Phaseolus polystachios
Phlox nivalis
Phoenix dactylifera (cultivated)
Phyllanthus carolinensis
Pinus clausa (?)
Pleopeltis polypodioides
Psilotum nudum
Psychotria nervosa
Rhizophora mangle
Rhynchospora ciliaris
Rivina humilis
Sagittaria subulata
Samolus ebracteatus
Schoenocaulon dubium
Sida rhombifolia

Sophora tomentosa
Sphenopholis obtusata
Stillingia sylvatica
Tillandsia recurvata
Tillandsia utriculata
Vallisneria americana
Viburnum obovatum
Vittaria lineata

The mosses appended here were described in the *Flora* and occur in the areas of Florida traveled by Michaux. To our knowledge, he did not collect any of these plants in Florida. Current names of the mosses were kindly provided by M. Crosby (pers. comm., January 29, 2000).

Dicranum glaucum var. *pumilum* Michx.=*Leucobryum albidum* (Brid.) Lindb.
Fissidens subbasilaris Hedwig
Funaria flavicans Michx.
Gymnostomum turbinatum Michx.=*Physcomitrium pyriforme* (Hedwig) De Not.
Hypnum hians Hedwig=*Eurhynchium hians* (Hedwig) Jaeg. & Sauerb.
Hypnum minutulum Hedwig=*Cyrto-hypnum minutulum* (Hedwig) W. R. Buck & H. Crum
Leskea adnata Michx.=*Sematophyllum adnatum* (Michx.) E. G. Britt.
Leskea rostrata Hedwig=*Anomodon rostratus* (Hedwig) Schimp.
Mnium cuspidatum Hedwig=*Plagiomnium cuspidatum* (Hedwig) T. Koponen
Mnium palustre Hedwig=*Aulacomnium palustre* (Hedwig) Schwaegr.
Neckera seductrix Hedwig=*Entodon seductrix* (Hedwig) C.M.
Polytrichum brachyphyllum Michx.=*Pogonatum brachyphyllum* (Michx.) P. Beauv.
Polytrichum perigoniale Michx.=*Polytrichum commune* Hedwig
Pterigynandrum hirtellum Hedwig=*Thelia hirtella* (Hedwig) Sull.
Pterigynandrum julaceum Hedwig=*Leucodon julaceus* (Hedwig) Sull.
Pterigynandrum trichomitriun Hedwig=*Forsstroemia trichomitria* (Hedwig) Lindb.
Trematodon longicollis Michx.
Trichostomum pallidum Hedwig=*Ditrichum pallidum* (Hedwig) Hampe
Weissia micro-odonta Hedwig=*Weissia controversa* Hedwig

Appendix 1–2

Plants Indicated by André Michaux to Be Part of the Florida Flora

This listing includes all species of plants that Michaux indicated in his *Flora* as occurring in Florida. In addition, we have added Florida plants not mentioned in the *Flora*, but found in the IDC 6211 microfiche collection, Michaux's *Journal*, primary literature, and letters, seed lists, and plant lists that he sent to France.

With few exceptions, the species are listed in the same order as they appear in the *Flora*. Original names, spellings, and descriptions in Latin are retained as given in the *Flora*. Family names have been added. Capitalized names are those in current use and follow Wunderlin (1998) and Radford et al. (1968). The IDC number, if present, corresponds to the IDC 6211 microfiche of the Michaux herbarium located in the Muséum National d'Histoire Naturelle at Paris. We have followed Uttal's (1984) system to quickly locate a specimen on a given microfiche card. In the example, IDC 2-8, the first number represents the microfiche card and the second number is the specific plant, reading left to right. Data on the herbarium sheets as originally written, including scientific names of the plants, place names, and all underlines have been retained. The herbarium labels are written in French or Latin. Punctuation marks have been added for clarification. Species in the *Flora* that lack a Florida locality, but are listed in the IDC with a Florida distribution, are designated with (**). Plants not in the *Flora* but with a Florida distribution based on the IDC and *Journal* are designated with (*). Florida plants whose scientific names were authored by Michaux and have stood the test of time are designated with (***). Three genera named by Michaux were represented by plants that he found while in Florida: *Ceratiola* Michx., *Arundinaria* Michx., and *Ipomopsis* Michx. The Florida rosemary, *Ceratiola ericoides* Michx., is the only endemic species represented in the listing below.

Boreali 1

ACANTHACEAE

Justicia humilis Michx.: *HAB.* in sylvarum paludibus caespitosis, a Carolina ad Floridam. (Affinis J. *assurgenti*, Linn.). JUSTICIA OVATA (Walt.) Lindau IDC 2-8, 9 (Justicia humilis; no locality given).

LAMIACEAE

Salvia coccinea L.: *HAB.* juxta fluvios et in insulis Floridae. SALVIA COCCINEA Buc'hoz ex Etl. IDC 4-3 (Salvia coccinea; in Maritimis Georgiae, Floridae).

Monarda lutea Michx.: *HAB.* a Virginia ad Floridam. (M. *punctata*. Linn.). MONARDA PUNCTATA L. IDC 4-12 (Monarda lutea, M. punctata L.; a Virginia ad Floridam prasertim in Maritimis).

HAEMODORACEAE

Heritiera gmelini Michx.: *HAB.* in paludibus et uliginosis a Carolina maritima ad Floridam. Junio florens. (*H. tinctorum*. Bull. des sc. n°. 19.). Tab. 4 of the *Flora* is this species. LACHNANTHES CAROLIANA (Lam.) Dandy IDC 5-21 (Heritiera gmelini; no locality given). Type (Uttal 1984).

XYRIDACEAE

(***) Xyris jupicai Michx.: *HAB.* a Marilandia ad Floridam. XYRIS JUPICAI Rich. IDC 6-6 (a Maryland ad Floridam).

IRIDACEAE

(*) Ixia caelestina Bartram pag. [unclear] Floride. CALYDOREA CAELESTINA (W. Bartr.) Goldblatt & Henrich IDC 6-8. Not in the *Flora*; IDC collection.

MAYACACEAE

Mayaca aubletii Michx.: *HAB.* a Carolina ad Floridam. MAYACA FLUVIATILIS Aubl. IDC 6-17 (Mayaca aubletii; no locality given).

CYPERACEAE

Cyperus hydra Michx.: *HAB.* in cultis Virginiae, Carolinae, Floridae. CYPERUS ROTUNDUS L. IDC 7-3 (Cyperus hydra; Hab. in Virginia maritima Carolina Georgia; another label reads: 1794 Caroline). Type (Uttal 1984).

Scirpus castaneus Michx.: *HAB.* in Florida. FIMBRISTYLIS SPADICEA (L.) Vahl IDC 8-5 (Scirpus castaneus; Hab. in Florida juxta Tomoko creek [Tomoka River, Volusia Co., Fla.]; another label reads Fimbristyles Novum genus ex Rich. A third label reads Scirpus castaneus Hab. in Carolina [sic Florida], juxta Tomoko creek). Type (Uttal 1984).

Scirpus spathaceus Michx.: *HAB.* in paludosis sylvaticis a sinu *Hudsonis* ad Floridam. (Cyperus *spathaceus*. Willd. *Spec, pl. 289*). DULICHIUM ARUNDINACEUM (L.) Britt. IDC 7-21 (Scirpus spathaceus; haute des terres et Territoire de la Baye d'Hudson).

(***) Schoenus ciliaris Michx.: *HAB.* in Florida. RHYNCHOSPORA CILIARIS (Michx.) C. Mohr IDC 9-10 (Schoenus ciliaris; Hab. in Florida, juxta Tomoko Creek legi). Type (Uttal 1984).

(***) Fuirena scirpoidea Michx.: *HAB.* in paludosis aestate exsiccabilibus Floridae. Tab. 7 of the *Flora* is of this species. FUIRENA SCIRPOIDEA Michx. IDC 10-9 (? Fuirena scirpoidea; Le 11 April vers le haut de Tomoko riv. [Scirpus articulatus is crossed out.]). Type (Uttal 1984).

POACEAE

Trichodium laxiflorum Michx.: *HAB.* in humidis et pratensibus a sinu *Hudsonis* ad Floridam (Cornucopiae hyemalis. Walt.). Tab. 8 of the *Flora* is this species. AGROSTIS HYEMALIS (Walt.) Britton et al. IDC 11-5 (Trichodium laxiflorum. Cornucopiae hyemais Walt. Hab. in pratensibus apricis a Canada ad Floridam). Hitchcock annotated this as *Agrostis scabra* (Michx.) Willd. Type (Uttal 1984).

Trichodium decumbens Michx.: *HAB.* a Virginia maritima ad Floridam, praesertim ad ripas amnium, solo limoso hieme inundato. (Cornucopiae perennans. Walt.). AGROSTIS PERENNANS (Walt.) Tuck. IDC 11-7 (Trichodium decumbens; Hab. in Carolina, praesertim in umbrosis ripariis Amnium. Trichod. (certissime) majus Cornucopiae perennans Walt.). Type (Uttal 1984).

Paspalum precox [praecox] Walt.: *HAB.* a Carolina ad Floridam. PASPALUM PRAECOX Walt. IDC 13-5 (Paspalum praecox; Hab a Carolina ad Floridam; on another label: in Florida).

(***) Paspalum floridanum Michx.: *HAB.* in Florida et Georgia. (*Obs.* Affine P. *virgato*. Linn.) PASPALUM FLORIDANUM Michx. IDC 13-6 (Paspalum floridanum; No. 5 Georgie et Floride). Type (Uttal 1984).

(***) Paspalum plicatulum Michx.: *HAB.* in Georgia et Florida.

PASPALUM PLICATULUM Michx. IDC 13-3 (Paspalum plicatulum; Hab in Georgia, Florida No 2). Type (Uttal 1984).

Digitaria sanguinalis Michx.: *HAB.* a Pensylvania ad Carolinam; in cultis.: in Florida maritima. (Panicum *sanguinale.* Linn.; Syntherisma *precox* [sic *praecox*]. Walt.). DIGITARIA CILIARIS (Retz.) Koel. IDC 11-14 (Digitaria sanguinalis; Hab. a Pensilvania ad Carolinam).

Panicum hirtellum L.: *HAB.* in umbrosis a Carolina ad Floridam. OPLISMENUS HIRTELLUS (L.) P. Beauv. IDC 12-3 (Panicum hirtellum; Hab in umbrosis sylvarum a Carolina maritima ad Floridam No. 13).

Panicum molle Michx.: *HAB.* in cespitosis Floridae. (*Obs.* Habitus Milii *punctati.* Linn.). ERIOCHLOA MICHAUXII (Poir.) Hitchc. IDC 12-11 (Panicum molle; Lieux très humides a 15 Miles [probably south] de St. Augustin. Another label reads Panicum molle; Hab. in sabulosis maritimis Floridae). Type (Uttal 1984).

Panicum virgatum L.: *HAB.* ad ripas fluviorum *Mississipi*, etc. a Virginia ad Floridam. (Panicum *coloratum*? Walt.). PANICUM VIRGATUM L. IDC 12-9 (Panicum virgatum; Hab. a Pensylvania ad Georgiam. Another label reads Pres le Débarquement du vieux . . . sw Cooper River Carolina; Voyes l'herbier des Illinois; Rare en Basse Caroline).

Agrostis indica Sw.: *HAB.* a Carolina ad Floridam. SPOROBOLUS INDICUS (L.) R. Br. IDC 13-14 (Agrostis indica; Florida, Carolines lieux aride; Another label reads Agrostis indica; Hab a Virginia maritima ad Floridam).

Erianthus saccharoides Michx.: *HAB.* a Carolina ad Floridam, in humidis. (Anthoxantum *giganteum.* Walt.). SACCHARUM GIGANTEUM (Walt.) Pers. IDC 13-19, 20 (Erianthus saccharoides; no locality given). IDC 13-19. Type (Uttal 1984).

Andropogon macrourum Michx.: *HAB.* a Carolina ad Floridam, in humidis. (Cinna *glomerata.* Walt.). ANDROPOGON GLOMERATUS (Walt.) Britton et al. IDC 14-1 (Andropogon macrourum; Hab. a Virginia ad Carolinam). Type (Uttal 1984).

Andropogon dissitiflorum Michx.: *HAB.* a Carolina ad Floridam, in sylvis (Cinna *lateralis.* Walt.). ANDROPOGON VIRGINICUS L. IDC 14-5 (Andropogon dissitiflorum; Hab. in Carolina, Georgia, Florida). Type (Uttal 1984).

(**), (***) Andropogon ternarium Michx.: *HAB.* in montosis Carolinae. ANDROPOGON TERNARIUS Michx. IDC 14-8 (Andropogon ternarium? No 68 Floride; another label gives: Andropogon ternarium; Hab. in regione Wabash, Georgia montosa etc; a third label gives Andropogon ternarium? Wabash, Illinois). Type (Uttal 1984).

Chloris petraea Sw.: *HAB.* in sabulosis maritimis Georgiae et Floridae. EUSTACHYS PETRAEA (Sw.) Desv. IDC 14-11 (Chloris petraea; Carolines et Florides).

(***) Tripsacum cylindricum Michx.: *HAB.* in sabulosis Floridae. COELORACHIS CYLINDRICA (Michx.) Nash IDC 14-17 (Tripsacum cylindricum; in Florida). Type (Uttal 1984).

Rottboella dimidiata L.: *HAB.* in maritimis, a Carolina and Floridam. STENOTAPHRUM SECUNDATUM (Walt.) Kuntze IDC 14-18 (Rottbola complaneta is crossed out on label; in maritimis a Carolina ad Florida).

Cenchrus tribuloides L.: *HAB.* a Virginia maritima ad Floridam. CENCHRUS TRIBULOIDES L. IDC 14-20 (Cenchrus tribuloides; no locality given).

(***) Aira obtusata Michx.: *HAB.* in aridis, a Carolina ad Floridam. SPHENOPHOLIS OBTUSATA (Michx.) Scribn. IDC 15-4 (Aira obtusata; Hab. in sabulosis Carolinae, Georgiae, Floridae. In Florida juxta domum Wiggin). Type (Uttal 1984).

Melica glabra Michx.: *HAB.* a Virginia ad Floridam. (*Obs.* Forsan huc referendae sunt M. *altissima* et *mutica.* Walt.). MELICA MUTICA Walt. IDC 15-5 (Melica glabra; a Carolina ad Floridam. Another label gives Melica glabra; No. 5 Florida f Matança, No. 5). Type (Uttal 1984).

Trachynotia polystachya Michx.: *HAB.* in inundatis maritimis, a Nova Anglia ad Floridam. SPARTINA CYNOSUROIDES (L.) Roth IDC 15-6 (Trachinotia (a dorso valvarium scabro) Dactylis cynosuroides; another label reads Dactylis L. polystachia, Trachinotia polystachia; Basse Caroline). Type (Uttal 1984).

Eleusine indica Michx.: *HAB.* in cultis, a Carolina ad Floridam. (Cynosurus *indicus.* Linn.). ELEUSINE INDICA (L.) Gaertn. IDC 15- 12 (Cynosarus indicus L. Eleusina gaertnea indica in cultis a Carolina ad Florida; another label reads Cynosarus indicus, Eleusine gaertn. Illinois).

Arundinaria macrosperma Michx.: *HAB.* ad ripas flum. *Mississipi*: in Carolina, Florida, etc. Martio flerens [florens]. ARUNDINARIA GIGANTEA (Walt.) Walt. ex Muhl. IDC 17-12 (Arundinaria macrosperma; Hab. a Virginia ad Floridam et in occidentalibus juxta fluvis ab Illinoensibus ad ostium Misissipi.). *Journal* plant.

RUBIACEAE

(***) Rubia Brownei Michx.: *HAB.* in umbrosis, a Carolina ad Floridam. (Valantia *hypocarpia*. Linn.; Rubia *peregrina*. Walt.). GALIUM HISPIDULUM Michx. IDC 20-14 (Rubia Brownei; in umbrosis Sylvarum Carolinae). Type (Uttal 1984).

Houstonia linnaei Michx.: *HAB.* in maritimis arenosis, a Virginia ad Floridam. (Var. *minor* Michx.). HEDYOTIS CRASSIFOLIA Raf. IDC 21-10 (Houstonia linnaei; Basse Carolin.). Type (Uttal 1984).

Houstonia rotundifolia Michx.: *HAB.* in apricis submaritimis Floridae et Carolinae. Martio floret. HEDYOTIS PROCUMBENS (J. F. Gmel.) Fosberg IDC 21-13 (Houstonia alba rotundifolia [alba crossed out]; Voyes la Descript au supplemt des Pl. de la Floride No. 1.[er]). Type (Uttal 1984).

Houstonia angustifolia Michx.: *HAB.* in submaritimis Floridae. HEDYOTIS NIGRICANS (Lam.) Fosberg IDC 21-19 (Houstonia angustifolia; no locality given). Type (Uttal 1984).

Cephalanthus occidentalis L.: *HAB.* in paludosis, a Canada ad Floridam. CEPHALANTHUS OCCIDENTALIS L. IDC 22-1 (Cephalanthus occidentalis; a Nova Anglia ad Floridam).

(*) Coffea Michx.: PSYCHOTRIA NERVOSA Sw. Not in the *Flora*. Plant mentioned in the *Journal* and listed in shipments of seeds from Florida sent to d'Angiviller on August 2, 1788, and to Thouin on January 5, 1789. S. Barrier (pers. comm.), of the Muséum National d'Histoire Naturelle, wrote in a letter dated July 20, 1994, that a specimen is in the Richard herbarium. Locality given is Florida.

PRIMULACEAE

(*) Samolus [crossed out] Lysimachia affinis: No. 8, Sur la riv. Aisa hatcha en Floride le 2 Avril 1787 [sic 1788]. Samodia ebracteata Baudo. SAMOLUS EBRACTEATUS Kunth IDC 27-6. Not in the *Flora*. This species is mounted next to *Samolus valerandi* L. which is in Boreali 1 of the *Flora*, but without a locality. *S. ebracteatus* was thought by Michaux to be a new *Spigelia*, as written on April 2, 1788, in his *Journal*.

SOLANACEAE

Lycium carolinianum Walt.: *HAB.* ad littora scirposa Carolinae, Georgiae, Floridae. (L. *salsum*. Bartram *trav*. 5g.). LYCIUM CAROLINIANUM Walt. IDC 24-20 (Lycium tetrandrum; in maritimis Georgiae).

RUBIACEAE

Chiococca racemosa L.: *HAB.* in maritimis Floridae. CHIOCOCCA ALBA (L.) Hitchc. IDC 26-4 (Chiococca racemosa; no locality given). Listed in shipments of seeds from Florida sent to d'Angiviller on August 2, 1788, and to Thouin on January 5, 1789.

ASCLEPIADACEAE

(*) Cynanchum. Georgie et Floride. CYNANCHUM. IDC 29-9 (listed on label as: 1. Cynanchum angulosum, 2. Cyn. muricatum, 3. Cyn. integrum, 4. Cyn. suberosum, and 5. Cyn. linearifolium). The IDC 29-9 specimen appears to be *Cynanchum scoparium* Nutt., a frequent species found in hammocks of the peninsula. Not in the *Flora*; IDC collection only.

(*) Asclepias shrub found on the St. Johns River, as mentioned in the *Journal* on May 8, 1788. The specific identity of this plant has not been determined.

LOGANIACEAE

Gelsemium nitidum Michx.: *HAB.* in Carolina inferiore, Georgia, Florida, Virginia maritima. GELSEMIUM SEMPERVIRENS (L.) W. T. Ait. IDC 29-11 (Carol; Bignonia sempervirens). IDC 29-12 (no locality given. No. 7 Bignonia sempervirens; Gelsemium nitidum). *Journal* species. Type (Uttal 1984).

CONVOLVULACEAE

Convolvulus obtusilobus Michx.: *HAB.* in littoribus arenosis Georgiae et Floridae. IPOMOEA IMPERATI (Vahl) Grisebach IDC 32-18 (Convolvulus *obtusilobus*; In littorib. arenosis Georgiae). Type (Uttal 1984).

Convolvulus dissectus L.: *HAB.* in collibus calcariis Floridae. MERREMIA DISSECTA (Jacq.) Hallier f. IDC 32-19 (Convolvulus dissectus? Floride). *Journal* plant.

Ipomoea lacunosa L.: *HAB.* in Florida et Carolina. (Affinis *I. trilobae.* Linn.). IPOMOEA LACUNOSA L. IDC 33-10 (Ipomoea lacunosa Lin. excluso Raji Synonymo.; de Floride).

Ipomoea bona nox L.: *HAB.* in sylvis ripariis Floridae. IPOMOEA ALBA L. IDC 33-17 (Ipomoea bona nox; in Florida). *Journal* plant.

(***) Ipomoea macrorhiza Michx.: *HAB.* in maritimis Georgiae et Floridae. IPOMOEA MACRORHIZA Michx. IDC 33-13 (Ipomoea macrorhiza; In maritimis Georgiae et Floridae). Type (Uttal 1984).

POLEMONIACEAE

(**) Ipomopsis elegans Michx.: *HAB.* in Carolina (Ipomoea *rubra.* Linn.; Cantua *coronopifolia.* Willd. *spec.* 879.). IPOMOPSIS RUBRA (L.) Wherry IDC 33-21 (Ipomaxtra cantua Jussieu Floride). In a published paper on some rare and new plants of North America, written in 1792 by André Michaux and edited by Lamarck, Michaux cited this species as "Ipomaea erecta, foliis pinnatifidis. Se trouve sur les bords de la mer, en Floride" (Rehder 1923).

(**) Phlox subulata L.: *HAB.* in aridis sabulosis Carolinae. PHLOX NIVALIS Lodd ex Sweet IDC 34-8 (Phlox subulata varieté trouvée en Floride). IDC specimen appears to be *Phlox nivalis* which occurs in Florida, but is labeled *Phlox subulata* which today does not occur in the state.

SOLANACEAE

(**) Solanum carolinense L.: *HAB.* in agris Carolinae. SOLANUM CAROLINENSE L. IDC 36-3 (Solanum carolinense; in Carolina, Georgia, in Florida).

RHAMNACEAE

(**) Rhamnus volubilis DC.: *HAB.* in Ludovisiana et Carolina. BERCHEMIA SCANDENS (Hill) K. Koch IDC 36-21 (Rhamnus volubilis; no locality given). Cited in the *Journal* as *Ziziphus scandens.*

(***) Rhamnus minutiflorus Michx.: *HAB.* in maritimis, a Carolina septentrionali ad Floridam. SAGERETIA MINUTIFLORA (Michx.) C. Mohr IDC 36-20 (Rhamnus *minutiflorus*; Hab. in maritimis a Carolina septentrionale ad Floridam). Type (Uttal 1984).

Ceanothus americanus L.: *HAB.* ubique a Canada ad Floridam. CEANOTHUS AMERICANUS L. IDC 37-6 (Ceanothus; no locality given).

(***) Ceanothus microphyllus Michx.: *HAB.* in herbosis sabulosis sylvarum Georgiae et Floridae. CEANOTHUS MICROPHYLLUS Michx. IDC 37-5 (No. 18 Ceanothus (floridanus). Le 16 Mars lieux arides vers Nord West river [Pellicer Creek, St. Johns Co., Fla.]). Shrub No. 18 mentioned in the *Journal.* Seeds of *Ceanothus* (*floridanus*) collected in Florida were sent to d'Angiviller on August 2, 1788. Then on January 5, 1789, seeds of *Ceanothus* nova species collected from Florida were sent to Thouin. Type (Uttal 1984).

APIACEAE

Eryngium foetidum L.: *HAB.* in campestribus, pratensibus aridis Antillarum et Floridae. (*Obs.* Excluso Gronovii synonymo.). ERYNGIUM FOETIDUM L. IDC 38-6 (In campestribus pratensibus aridis Antillarum). Plant is not native to the U.S.

(**), (***) Eryngium yuccifolium Michx.: *HAB.* in paludosis Virginiae. ERYNGIUM YUCCIFOLIUM Michx. IDC 38-7 (Eryngium yuccifolium; In apricis Sylvarum inter gramineas a Virginia ad Floridam). Type (Uttal 1984).

GENTIANACEAE

Gentiana saponaria L.: *HAB.* in herbosis sylvarum, a Canada ad Floridam. GENTIANA SAPONARIA L. IDC 40-7 (Gentiana saponaria; No. 1.[er]; no locality given). Species is uncommon today in the Florida panhandle.

CAPRIFOLIACEAE

(**) Viburnum cassinoides L.: *HAB.* in umbrosis sylvarum, justa rivos Carolinae inferioris et Georgiae. VIBURNUM OBOVATUM Walt. IDC 41-4 (Viburnum cassinoides; in umbrosis ad ripas rivulorum Carolinae Georgiae). *Viburnum cassinoides* is listed in Michaux's *Journal*. The herbarium specimen labeled *V. cassinoides*, however, is *V. obovatum* Walt.

AMARYLLIDACEAE

(**) Pancratium mexicanum L.: *HAB.* in inundatis Carolinae. HYMENOCALLIS LATIFOLIA (Mill.) M. Roem. IDC 43-20 (Pancratium mexicanum; no locality given). *Journal* plant.

(*) Crinum americanum L. CRINUM AMERICANUM L. Not in the *Flora*; *Journal* plant.

JUNCACEAE

(*) Throughout his Florida account, Michaux wrote of unspecified species of rushes [joncs]. The black needlerush, *Juncus roemerianus* Scheele, is an excellent candidate for the juncus he probably observed.

COMMELINACEAE

(**) Tradescantia rosea Vent.: *HAB.* in subudis sabulosis Carolinae. CALLISIA ROSEA (Vent.) D. R. Hunt IDC 44-15 (Tradescantia rosea; a Carolina ad Floridam).

PONTEDERIACEAE

Pontederia cordata L.: *HAB.* in aquosis, a Canada ad Floridam. PONTEDERIA CORDATA L. IDC 44-17 (Pontederia cordata; a Canada ad Floridam).

ARACEAE

Orontium aquaticum L.: *HAB.* in aquosis, a Canada ad Floridam. ORONTIUM AQUATICUM L. IDC 45-1 (Orontium aquaticum; a Nova Anglia ad Carolinam; another label reads Orontium de N. York).

BROMELIACEAE

Tilliandsia recurvata L.: *HAB.* in Florida. TILLANDSIA RECURVATA (L.) L. IDC 45-7 (Tilliandsia recurvata; in Florida).

(**) Tilliandsia usneoides L.: *HAB.* in arboribus littoralibus Carolinae. TILLANDSIA USNEOIDES (L.) L. IDC 45-5 (Tilliandsia *usneoides* in maritimis a Carolina ad Floridam). In an archival box containing Michaux's cahiers located at the Library of the American Philosophical Society in Philadelphia, there is a loose sheet entitled, "Notes of topographic botany." Michaux wrote, "p. 87 Tillandsia usneoides depuis le 35 deg. de lat. +. jusqu' a de la du 28."

(*) Tillandsia: in Florida arboribus Adelia porulosa legi. TILLANDSIA UTRICULATA L. IDC 45-8. Not in the *Flora*. Seeds collected from Florida of the synonym, Tillandsia lingulata, were sent to Thouin on 5 January 1789.

AGAVACEAE

Yucca aloifolia L.: *HAB.* in littoralibus Carolinae, Floridae, &c. YUCCA ALOIFOLIA L. IDC 45-10 (Yucca aloifolia; no locality given).

ARECACEAE

Chamaerops Palmetto Michx.: *HAB.* in maritimis, a Carolina ad Floridam. (Corypha *Palmetto*. Walt.). SABAL PALMETTO (Walt.) Lodd. ex Schult. & Schult. f. IDC 47-8 (Chamaerops Palmetto; no locality given). *Journal* plant.

Chamaerops serrulata Michx.: *HAB.* in maritimis Georgiae Floridaeque. SERENOA REPENS (W. Bartr.) Small IDC 47-10 (Environs de Sunbury en Georgia et Savanah; Entre Savanah et Sunbury). *Journal* plant. Type (Uttal 1984).

LILIACEAE

(***) Helonias dubia Michx.: *HAB.* in sabulosis Georgiae et Floridae. SCHOENOCAULON DUBIUM (Michx.) Small IDC 48-1 (Helonias dubia; Floride). Type (Uttal 1984).

ALISMATACEAE

Alisma subulata L.: *HAB.* in Florida. (affinis *A. ranunculoidi.*). SAGITTARIA SUBULATA (L.) Buchenau IDC 48-20 (Alisma subulata; Georgia, Florida).

MELASTOMATACEAE

Rhexia lutea Walt.: *HAB.* in Florida et Georgia. RHEXIA LUTEA Walt. IDC 49-15 (Rhexia hispida, flore luteo; no locality given).

ERICACEAE

Vaccinium stamineum L.: *HAB.* a Pensylvania ad Floridam. VACCINIUM STAMINEUM L. IDC 53-4 (No. 3A Georgie); IDC 53-5 (Vaccinium stamineo affine; Georgie, Florida).

Vaccinium frondosum L.: *HAB.* in pinetis, aridis, a Virginia ad Floridam. GAYLUSSACIA DUMOSA (Andrews) T. & G. IDC 52-18 (Vaccinium frondosum Linn. Note on label by Asa Gray gives species as Vaccinium dumosum).

Vaccinium arboreum Marsh.: *HAB.* in sylvis subaridis, a Carolina ad Floridam. VACCINIUM ARBOREUM Marsh. IDC 52-15 (Vaccinium arboreum diffusum; Hort. Kew. IDC 52-16 (No 2, Vaccinium arboreum).

Vaccinium myrsinites Lam.: *HAB.* in sabulosis aridis Floridae. VACCINIUM MYRSINITES Lam. IDC 52-7 (No 1er Vaccinium myrsinites A. ovale); IDC 52-8 (Vaccinium myrsinites α. ovale. Vaccinium myrsinites β. lanceolatum or V. tenellum Swamps en Georgie).

POLYGONACEAE

Polygonum virginianum Michx.: *HAB.* in sylvis umbrosis, a Canada ad Floridam. POLYGONUM VIRGINIANUM L. IDC 53–13 (Polygonum virginian. Bois humides de l'Am. Sept., Kentuckey).

Polygonum sagittatum L.: *HAB.* in humidis apertis, herbosis, a Carolina ad Floridam. POLYGONUM SAGITTATUM L. IDC 54–15 (Polygonum sagittatum; Canada, Pensylvania, Carolines).

SAPINDACEAE

(**) Sapindus saponaria L.: *HAB.* juxta littora, in Georgia. SAPINDUS SAPONARIA L. IDC 54–19 (Sapindus saponaria; no locality given). *Journal* plant.

LAURACEAE

Laurus pseudo-Benzoin Michx.: *HAB.* juxta rivulos et in udis, a Canada ad Floridam. (*L. Benzoin* Linn. ex synon. Pluckneti.). LINDERA BENZOIN (L.) Blume IDC 55–4 (Laurus pseudobenzoin; Pres d'Alexandrie en Virginie). Type (Uttal 1984).

Laurus sassafras L.: *HAB.* a Canada ad Floridam. SASSAFRAS ALBIDUM (Nutt.) Nees IDC 55–7 (Larus sassafras; Riv Savana, Riv S[te] Maria). IDC 55–8 (L. sassafras; no locality given).

Laurus catesbyana Michx.: *HAB.* in Florida calidiore et *Bahama.* OCOTEA CORIACEA (Sw.) Britt. Plant not seen in IDC collection. Michaux s.n. in Herb. Richard (holotype, P). Cited in Rohwer (1993). In a published paper on some rare and new plants of North America, written in 1792 by André Michaux and edited by Lamarck, *Laurus indica* is listed. *L. indica* is *Ocotea coriacea.* Michaux wrote concerning the species: "Ne se trouve qu' aux isles Bahame. Régions meridionales de la Florida." (Rehder 1923).

(**) Laurus caroliniensis Catesb.: *HAB.* in Carolina, Ludovisia. (*Obs.* Perperam a Linnaeo ad *L. Borboniam.* relata.). PERSEA BORBONIA (L.) Spreng. IDC 55-14 (Laurus caroliniensis borbonia; no locality given). 55-16 (Laurus borbonia, des Jardiniers Anglais, Laurus caroliniensis). 55-17 (No 5g; Laurus *caroliniensis*; no locality given). 55-18 (Laurus *caroliniensis*; Georgie). *Journal* plant.

ERICACEAE

Andromeda ferruginea Walt.: *HAB.* in Florida et Georgia. LYONIA FERRUGINEA (Walt.) Nutt. IDC 57-11 (no 3. Andromeda folius revolute subt—ferruginis; St. Augustine). *Journal* plant.

Andromeda nitida Bartr.: *HAB.* in Carolina inferiore et Florida. (*A. lucida* Lam. *Dict.* n°. 9. & *A. coriacea Hort. Kew.*). LYONIA LUCIDA (Lam.) K. Koch IDC 57-4 (Andromeda nitida; Caroline) and (***) LYONIA FRUTICOSA (Michx.) G. S. Torr. IDC 57-1 (Andromeda coriacea?; Florida).

Andromeda laurina Michx.: *HAB.* in Florida. (*A. formosissima* Bart., *A. reticulata* Walt., *A. populifolia* Lam. Dict. *A. acuminata* Ait., *A. lucida* Jacq.). AGARISTA POPULIFOLIA (Lam.) Judd IDC 57-10 (Andromeda formosissima de Bartram. Andromeda foliis reticulatis. No 19 Andromeda, 17 & 18 Mars, Andr. reticulata). *Journal* plant.

(**) Andromeda paniculata L.: *HAB.* in frigidioribus, per *Etats-Unis* (Var. *nudiflora*); in sylvis Carolinae inferioris; in stagnosis (Var. *foliosiflora*). LYONIA LIGUSTRINA (L.) DC. IDC 56-13 (Andromeda paniculata; Floride); IDC 56-16 (Floride). Both specimens are of the variety *foliosiflora*.

Andromeda arborea L.: *HAB.* in montibus *Alleghanis*, a Pensylvania ad Floridam. OXYDENDRUM ARBOREUM (L.) DC. IDC 58-1 (And arborea; no locality given). *Journal* plant. Plant occurs today in the panhandle of Florida.

Andromeda racemosa L.: *HAB.* a Pensylvania ad Floridam. (A. *paniculata*. Walt.). LEUCOTHOE RACEMOSA (L.) A. Gray IDC 57-20 (Andromeda racemosa, Floride; another label gives Pensylv. et N. Jersey, Andromeda racemosa).

Andromeda mariana L.: *HAB.* a Pensylvania ad Floridam. LYONIA MARIANA (L.) D. Don IDC 57-14 (Andromeda mariana; Floride).

(**) Kalmia hirsuta Walt.: *HAB.* in sylvis Georgiae. (*K. ciliata*, Bartram's *Travels*, pag. 18.). KALMIA HIRSUTA Walt. IDC 58-18 (Kalmia hirsuta; Floride et Georgie 69). In a published paper on some rare and new plants of North America, written in 1792 by André Michaux and edited by Lamarck, Michaux remarked on this species "On le trouve dans la Georgie et la Floride" (Rehder 1923).

Befaria paniculata Michx.: *HAB.* in Floridae arenosis. Tab. 26 of the *Flora* is of this species. BEJARIA RACEMOSA Vent. IDC 63-11 (no data). IDC 63-13 (Befaria (viscosa); Partie méridionale de la Georgie et Floride). Type (Uttal 1984).

FABACEAE

Podalyria tinctoria Michx.: *HAB.* a Canada ad Floridam. (Sophora *tinctoria*. Linn.). BAPTISIA LECONTII T. & G. IDC 61-1 (Sophora *tinctoria*; no locality given).

(*) Sophora occidentalis.: SOPHORA TOMENTOSA L. Not in the *Flora*; *Journal* plant. Species also listed in shipments of seeds from Florida sent to d'Angiviller on August 2, 1788, and to Thouin on January 5, 1789.

RHIZOPHORACEAE

(*) Rhizophora mangle L.: RHIZOPHORA MANGLE L. Not in the *Flora*;

Journal plant. Observed on Anastasia Island and in Mosquito Lagoon near Turtle Mound.

COMBRETACEAE

(*) Conocarpus racemosa Michx.: CONOCARPUS ERECTUS L. Not in the *Flora*. Species listed in shipments of seeds from Florida sent to d'Angiviller on August 2, 1788, and to Thouin on January 5, 1789. S. Barrier, Muséum National d'Histoire Naturelle, wrote in a letter dated July 20, 1994, that this species is in the Richard herbarium with the locality on the label as "Floridae littora."

(*) Mangrove with fruits like those of Catesby's fig tree. Michaux wrote this description in his *Journal* and this probably refers to the white mangrove, LAGUNCULARIA RACEMOSA (L.) C. F. Gaertn as shown in Catesby (1771). Species not in the *Flora*; *Journal* plant.

LYTHRACEAE

Lythrum verticillatum L.: *HAB.* in paludosis, a Canada ad Floridam. (Decodon. Gmel. *Syst.* 677.). DECODON VERTICILLATUS (L.) Ell. IDC 63-16 (Lythrum verticillatum; Caroline, Georgie, Pensylvanie no. 176 Walt.).

CHRYSOBALANACEAE

Chrysobalanus oblongifolius Michx.: *HAB.* in sabulosis sylvarum Georgiae et Floridae. LICANIA MICHAUXII Prance IDC 64-3 (Chrysobalanus oblongifolius; ad fluvium Ste *Mary* in Georgiae). Type (Uttal 1984).

ROSACEAE

Cerasus caroliniana Ait.: *HAB.* a Carolina ad Floridam. PRUNUS CAROLINIANA (Mill.) Ait. IDC 64-5 (Cerasus caroliniana, Padus sempervirens; Alatamaha).

Spiraea trifoliata L.: *HAB.* a Canada ad Floridam. GILLENIA TRIFOLIATA (L.) Moench IDC 66-18 (Spirea trifoliata on one label; another label gives Spiraea opulifolia; no locale given on either label). Species does not occur in Florida today.

Fragaria canadensis Michx.: *HAB.* in montosis et sylvis, a sinu *Hudsonis* ad Floridam. FRAGARIA VIRGINIANA Duchesne IDC 68-10 (Fragaria canadensis; Hab. ad Lacus Mistassinos). Rare today in the Florida panhandle.

RANUNCULACEAE

Actaea racemosa L.: *HAB.* a Canada ad Floridam, per tractus montium. CIMICIFUGA RACEMOSA (L.) Nutt. IDC 70-3 (*Actaea* racemosa; no locality given). Species does not occur in Florida today.

PAPAVERACEAE

Sanguinaria canadensis L.: *HAB.* a Canada ad Floridam. SANGUINARIA CANADENSIS L. IDC 70-7 (Affinis Sanguinaria sur les rives de l'Ohio). Species does not occur in Florida today where Michaux traveled.

SARRACENIACEAE

Sarracenia flava L.: *HAB.* in humidis apertis, a Carolina ad Floridam. SARRACENIA FLAVA L. IDC 70-15 (no locality given).

Sarracenia variolaris Michx.: *HAB.* a Carolina ad Floridam. SARRACENIA MINOR Walt. IDC 70-14 (no locale given). Type (Uttal 1984).

(***) Sarracenia psyttacina Michx.: *HAB.* ab urbe *Augusta* Georgiae ad Floridam. SARRACENIA PSITTACINA Michx. IDC 70-16 (Sarracenia psyttacina, no locality given). Species does not occur in Florida today where Michaux traveled.

NYCTAGINACEAE

(*) Pisonia inermis Jacq.. Not in the *Flora*. On March 9, 1794, Michaux wrote in his *Journal* that this shrub, with opposite leaves and branches, occurred along the coast of Carolina, Georgia, and Florida. None of the species of *Pisonia* occur today in Georgia and the Carolinas. *P. aculeata* L. and *P. rotundata* Griseb. do not occur today in Florida where Michaux traveled. *Guapira discolor* (Spreng.) Little (formerly *Pisonia discolor* Spreng.) occurs in Brevard County, where Michaux might have seen the plant.

PHYTOLACCACEAE

(*) Rivina humilis L.: RIVINA HUMILIS L. Not in the *Flora*; *Journal* plant.

ILLICIACEAE

Illicium floridanum L.: *HAB.* in Florida occidentali juxta flumen *Mississipi*. ILLICIUM FLORIDANUM J. Ellis Not in IDC material. Johann D. Schoepf wrote in 1784 that the species was found in the St. Augustine area, but that it was not so plentiful as in West Florida (Schoepf 1911).

(***) Illicium parviflorum Michx.: *HAB.* in Florida orientali, juxta amnem *S. Joannis* (Illicium *anisatum*. Bartram's *Journal*. pag. 24. Lond. 1769.). ILLICIUM PARVIFLORUM Michx. ex Vent. IDC 73-3 (Illicium floribus flavis; Floride). *Journal* plant. Type (Uttal 1984).

MAGNOLIACEAE

Liriodendron tulipifera L.: *HAB.* a Canada ad Virginiam et a Carolina ad Floridam, per tractus montium. (Var. *acutiloba*). LIRIODENDRON TULIPIFERA L. IDC 73-5 (Liriodendron tulipifera; varietes dans les Montagnes de la Caroline).

(**) Magnolia grandiflora L.: *HAB.* a finibus septentrionalibus Carolinae ad *Mississipi*. MAGNOLIA GRANDIFLORA L. IDC 73-12 (Magnolia grandiflora Floride). In an archival box containing Michaux's cahiers at the Library of the American Philosophical Society in Philadelphia, there is a loose sheet titled, "Notes of topographic botany." Michaux wrotes, p. 86, "the Magnolia grandiflora on the River St. Johns in Florida 100 pieds de haut et +." *Journal* plant.

Magnolia glauca L.: *HAB.* a Nova Caesarea ad Floridam. MAGNOLIA VIRGINIANA L. IDC 73-13 (Magnolia glauca; du Pensylvanie, Maryland, Virginie et Caroline Septentrionale). *Journal* plant.

ANNONACEAE

Orchidocarpum arietinum Michx.: *HAB.* a Virginia ad Floridam, juxta inundatas amnium ripas (Annona *triloba*. Linn.). ASIMINA TRILOBA (L.) Dunal IDC 74-2 (Orchidocarpon arietinum, Annona triloba; Pensylvania Virginie Carolines; Ohio et Misissipi [Rivers]). Type (Uttal 1984). Species does not occur in Florida today where Michaux traveled.

Orchidocarpum pygmaeum Michx.: *HAB.* in Georgia et Florida (Annona *pygmaea*. Bartr. *Trav.*). ASIMINA PYGMAEA (W. Bartr.) Dunal IDC 73-19 (Annona lanceolata pygmaea [lanceolata is crossed out]. No. 9 Georg-91).

Orchidocarpum grandiflorum Michx.: *HAB.* in Georgia et Florida (Annona *grandiflora*. Bartr. *Trav.* 8 or 18). ASIMINA OBOVATA (Willd.) Nash IDC 73-20 (No 8 Annona grandiflora, Orchidocarpum grandisflorum; Floride, Georgie). Kral (1960) concluded that this pawpaw of Michaux is *Asimina obovata*. Type (Uttal 1984). *Journal* plant.

(*) Annona glabra L. Bahamas. ANNONA GLABRA L. IDC 73-18. Not in the *Flora*; *Journal* plant. It is not sure that Michaux saw this plant in Florida.

Boreali 2

SCROPHULARIACEAE

Bartsia coccinea L.: *HAB.* a Nova Anglia ad Floridam. CASTILLEJA COCCINEA (L.) Sprengel IDC 78-7 (Bartsia coccinea; Connecticut, Canada, Pensylvanie, hautes Montagnes de Carolines, Georgie). Species does not occur in Florida today.

Gerardia flava L.: *HAB.* a Nova Anglia ad Floridam. AUREOLARIA VIRGINICA (L.) Pennell IDC 78-18 (Gerardia flava; no locality given). Species does not occur in Florida today where Michaux traveled.

Antirrhinum canadense L.: *HAB.* a Canada ad Floridam, in arvis. LINARIA CANADENSIS (L.) Chaz. IDC 78–20 (Antirrhinum canadense?; no locality given).

ACANTHACEAE

Ruellia humistrata Michx.: *HAB.* ad fines Georgiae et Floridae. RUELLIA CAROLINIENSIS (J. F. Gmel.) Steud. IDC 79-17 (Ruellia humistrata; sur la riv. Ste Marie en Georgie). Type (Uttal 1984).

Ruellia strepens L.: *HAB.* a Virginia ad Floridam. RUELLIA STREPENS L. IDC 79-13 (Ruellia strepens, Ruellia filamentis diadelphis; Caroline). Species does not occur in Florida today.

BIGNONIACEAE

Bignonia capreolata L.: *HAB.* a Virginia ad Floridam. BIGNONIA CAPREOLATA L. IDC 80-4 (no locality given).

Bignonia radicans L.: *HAB.* a Pensylvania ad Floridam. CAMPSIS RADICANS (L.) Seemann ex Bureau IDC 80-2 (Bignonia radicans; no locality given).

AMARANTHACEAE

Illecebrum vermiculatum L.: *HAB.* in maritimis Floridae. BLUTAPARON VERMICULARE (L.) Mears IDC 81-15 (Illecebrum vermiculatum; In Maritimis Floride).

PASSIFLORIACEAE

Passiflora incarnata L.: *HAB.* a Virginia ad Floridam. PASSIFLORA INCARNATA L. IDC 81-18 (Passiflora; no locality given).

Passiflora lutea L.: *HAB.* in Virginia et regione Illinoensi, usque ad Floridam. PASSIFLORA LUTEA L. IDC 81-21 (Virginia, Carolines).

CARIACEAE

(*) Carica papaya L.: CARICA PAPAYA L. Not in the *Flora*; *Journal* plant.

STYRACACEAE

Halesia parviflora Michx.: *HAB.* in Florida, circa *Matança.* HALESIA CAROLINA L. IDC 82-14 (Halesia a petit fruit; Floride). Species was also called *Halesia tetraptera* L. in the *Journal* and in the *Flora* (*HAB.* in Sylvis umbrosis et rivulosis Carolinae), but this is a synonym. Type (Uttal 1984).

SYMPLOCACEAE

Hopea tinctoria L.: *HAB.* a Carolina ad Floridam. (*Obs.* Genus Symploco valde affine!). SYMPLOCOS TINCTORIA (L.) L'Hér. IDC 82-20 (Hopea; commence en Virginie 2 mile avant de d'entrer dans la Carol. Sept. IDC 82-21 (Hopea tinctoria; no locality given).

THEACEAE and MALVACEAE

Gordonia lasianthus L.: *HAB.* in maritimis Carolinae et Floridae. GORDONIA LASIANTHUS (L.) Ellis IDC 83-4 (Gordonia lasianthus; In Maritimes Carolinae et Floridae, 38 miles au Nord de Wilmington). *Journal* plant.

Sida rhombifolia L.: *HAB.* in ruderatis Floridae. SIDA RHOMBIFOLIA L. IDC 83-11 (*rhombifolia*? Sida; de Floride dont l'extremité des tiges était gelé. Planta annua in Carolina, frutescens in Floride, Sida rhombifolia?).

Hibiscus scaber Michx.: *HAB.* in maritimis Carolinae et Floridae. (*H. aculeatus*? Walt.). HIBISCUS ACULEATUS Walt. IDC 84-3 (Hibiscus rugosus; no locality given).

(***) Hibiscus grandiflorus Michx.: *HAB.* in maritimis Georgiae et Floridae et in regione *Natchez* ad *Mississipi.* (*Obs.* Affinis *H. clypeato.*). HIBISCUS GRANDIFLORUS Michx. IDC 84-6 (Hibiscus grandiflorus; Lieux maritimes). Type (Uttal 1984).

Hibiscus speciosus Ait.: *HAB.* in Floridae scirposis. (*H. coccineus.* Bart.

Walt.). HIBISCUS COCCINEUS Walt. IDC 83-20 (Hibiscus speciosus Ait.; no locality given).

Malva caroliniana L.: *HAB.* in Carolina et Virginia. MODIOLA CAROLINIANA (L.) G. Don. IDC 83-13 (Malva carol; no locality given). Seeds collected in Florida were sent to Thouin on January 5, 1789. Not in the *Journal.*

Bombax gossypinum L. COCHLOSPERMUM RELIGIOSUM (L.) Alston. Seeds collected in Florida were sent to Thouin on January 5, 1789. Species neither in the *Flora* nor in the *Journal.*

FABACEAE

(***) Petalostemum carneum Michx.: *HAB.* in Georgia et Florida. DALEA CARNEA (Michx.) Poir. IDC 84-19 (*No. 17* Floride, Georgie, Kentucky, Psoraloides arbuste legumineux; Lieux humides pres Nord West riv. en Floride). Probably leguminous shrub No. 17 mentioned in the *Journal* and not *Erythrina herbacea* as annotated by Sargent (1889). Type (Uttal 1984).

POLYGALACEAE

Polygala lutea L.: *HAB.* a Marylandia ad Floridam. (Var. *elatior*) POLYGALA LUTEA L. IDC 85-15 (Polygala lutea; en Georgie).

Polygala corymbosa Michx.: *HAB.* in paludosis, a Carolina ad Floridam (P. *cymosa.* Walt.). POLYGALA CYMOSA Walt. IDC 85-6 (Polygala corymbosa minor; 2e Varieté Georgie 17-3). Type (Uttal 1984).

FABACEAE

Lupinus perennis L.: *HAB.* a Canada ad Floridam, in loci a mare remotis. LUPINUS PERENNIS L. IDC 85-19 (Lupinus perennis; sur la rivière Ste Marie).

Lupinus pilosus L.: *HAB.* a Carolina ad Floridam. LUPINUS VILLOSUS Willd. IDC 85-20 (Lupinus pilosus; Trouvé le 25 Avril 1787, Pres la riv. Savanah; a 3 milles de Alatamaha, route de Sunbury). Lupinus pilosus flore ceruleo in the *Journal* from Florida is probably *L. diffusus* Nutt. and not *L. villosus.*

(*) Lupinus pilosus flore ceruleo. LUPINUS DIFFUSUS Nutt. The description in the *Journal* indicates this species and not *L. villosus* Willd., which is listed in the *Flora* and collected in Georgia.

Psoralea melilotoides Michx.: *HAB.* in Carolina et Florida. (Trifolium psoralioides. Walt.). (*Obs.* Flores coerulei.). ORBEXILUM PEDUNCU-

LATUM (Mill.) Rydb. IDC 86-5 (Trifolium (psoralea) Walt. Psoralea melilotoides; no locality given). Type (Uttal 1984).

(**) Phaseolus paniculatus Michx.: *HAB.* in regione Illinoensi. PHASEOLUS POLYSTACHIOS (L.) Britton et al. IDC 86-19 (Phaseolus ou Dolichos non determiné; Phaseolus? sables arides non determiné). This species is probably "the Phaesol or Dolichos with big fruit" Michaux mentions in the *Journal.* According to Harper (1958) and the IDC image, the species is probably *Phaseolus polystachios.*

(*) Dolichos ensiformis Michx.: IDC 91-6 (Pois de sabre des Isles Antilles de l'Amerique. Le trouve aussi en Floride et en Georgia sur les bords de la Mer). CANAVALIA ENSIFORMIS (L.)DC. Synonym given in Sauer (1964). Species not in the *Flora.*

Erythrina herbacea L.: *HAB.* in Carolina et Florida. ERYTHRINA HERBACEA L. IDC 86-20 (no locality given). *Journal* plant. Seeds collected in Florida of *Erythrina arborea* were sent to Thouin on January 5, 1789.

Clitoria virginiana L.: *HAB.* in Virginia et Florida. (*Obs.* Valde affinis *C. brasilianae.*). CENTROSEMA VIRGINIANUM (L.) Benth. IDC 87-9 (Clitoria virginiana Michx. on the label but not in Michaux's handwriting; no locality given).

Amorpha fruticosa L.: *HAB.* in Carolina et Florida. AMORPHA FRUTICOSA L. IDC 88-6 (Amorpha fruticosa; no locale given). Genus, and probably this species, is listed in a shipment of seeds collected from Florida plants sent to d'Angiviller on August 2, 1788.

Galega virginiana L.: *HAB.* a Canada ad Floridam. (*Obs.* Affinis praesertim habitu *G. caribeae.*). TEPHROSIA VIRGINIANA (L.) Pers. IDC 88-17 (Galega virginiana; no locality given).

Galega villosa Michx.: *HAB.* a Carolina ad Floridam. TEPHROSIA SPICATA (Walt.) T. & G. IDC 88-19 (Galega villosa; Caroline et Floride). Type (Uttal 1984).

Indigofera caroliniana Walt.: *HAB.* a Carolina ad Floridam. INDIGOFERA CAROLINIANA Mill. IDC 89-4 (Indigofera perennis on label, but not in Michaux's handwriting; no locality given).

(***) Aeschynomene viscidula Michx.: *HAB.* in arenosis insulae *Cumberland* et Floridae. AESCHYNOMENE VISCIDULA Michx. IDC 90-21 (Aeschynomene viscidula; no locality given). Type (Uttal 1984).

Aeschynomene platycarpa Michx.: *HAB.* in humidis Carolinae, usque ad Floridam. SESBANIA VESICARIA (Jacq.) Ell. IDC 90-20 (Aeschynomene de Caroline, Georgie et de Floride. Another label gives Robinia

Jacq. vesicaria? ?Planta diadelpha in Florida? Aeschynomene platycarpa 1787).

(*) Guilandina bonducella Michx. CAESALPINIA BONDUC (L.) Roxb. Species listed in shipments of seeds from Florida sent to d'Angiviller on August 2, 1788, and to Thouin on January 5, 1789. Not in the *Flora*; *Journal* plant.

RUTACEAE

(*) Amyris elemifera L. Neither in the *Flora* nor in the IDC collection. Species is listed in a shipment of seeds from Florida sent to d'Angiviller on August 2, 1788. S. Barrier, of the Muséum National d'Histoire Naturelle, wrote in a letter dated July 20, 1994, that a Michaux specimen is in the Paris herbarium. Locality given as "Florida Cap Canaveral."

CLUSIACEAE

Ascyrum amplexicaule Michx.: *HAB.* in Florida. HYPERICUM TETRAPETALUM Lam. IDC 91-8 (Ascyrum amplexicaule; Floride).

Hypericum glaucum Michx.: *HAB.* in Florida. HYPERICUM MYRTIFOLIUM Lam. IDC 91-21 (Hypericum glaucum, Floridae). Type (Uttal 1984).

(**) Hypericum angulosum Michx.: *HAB.* in paludosis Carolinae. (*H. denticulatum*? Walt.). HYPERICUM DENTICULATUM Walt. IDC 92-6 (Hypericum angulosum, in humides a Carolina ad Floridam). Type (Uttal 1984). Species occurs in the Florida panhandle.

(**) Hypericum simplex Michx.: *HAB.* in Carolina inferiore. (Ascyrum *villosum*. Linn.; Hypericum *pilosum*. Walt.). HYPERICUM SETOSUM L. IDC 92-5 (Hypericum pilosum Walt., Hab. a Carolina ad Floridam, Ascyrum villosum L.).

ASTERACEAE

(**), (***) Lactuca graminifolia Michx.: *HAB.* in Carolina inferiore. LACTUCA GRAMINIFOLIA Michx. IDC 93-1 (Lactuca graminifolia; Floride et Basse [Caroline]). Type (Uttal 1984).

Hyoseris virginica L.: *HAB.* a Canada ad Floridam. (*Obs.* Habitus parvuli *Taraxaci.*). KRIGIA VIRGINICA (L.) Willd. IDC 93-11 (Hyoseris virginica?; no locality given).

(***) Liatris elegans Michx.: *HAB.* in Carolina et Florida (Staehelina *elegans* Walt.; Serratula *speciosa* Ait.). LIATRIS ELEGANS (Walt.) Michx. IDC 94-2 (no locality given).

Liatris paniculata Michx.: *HAB.* a Virginia ad Floridam. CARPHEPHORUS PANICULATUS (Gmel.) Herb. IDC 94-3 (Serratula glauca; Lieux humides dans les bois. Basse Caroline; Liatris paniculata).

Liatris odoratissima Michx.: *HAB.* a Carolina ad Floridam. CARPHEPHORUS ODORATISSIMUS (Gmel.) Herb. IDC 94-7 (Liatris odoratissima; no locality given).

Sparganophorus verticillatus Michx.: *HAB.* in inundatis, a Carolina ad Floridam. (Ethulia *uniflora*? Walt.). Tab. 42 of the *Flora* is of this species. SCLEROLEPIS UNIFLORA (Walter) Britton et al. IDC 95-3 (Sparganophorus verticillatus; Planta paludosa Caroline, Aethulia uniflora de Walter). Type (Uttal 1984).

Eupatorium connatum Michx.: *HAB.* a Canada ad Floridam (*E. perfoliatum* Linn.). EUPATORIUM PERFOLIATUM L. IDC 95-7 (Eupatorium perfoliatum; No 21, Carolines, Virginie, Illinois, Quebec). Type (Uttal 1984).

Eupatorium aromaticum L.: *HAB.* a Virginia ad Floridam. AGERATINA AROMATICA (L.) Spach IDC 95-18 (No. 19 Eupatorium aromaticum; habitat in Carolina, Basse Caroline).

Chrysocoma capillacea Michx.: *HAB.* in pascuis, juxta *Charlstown* et in Florida (*E. foeniculoides*? Walt.). EUPATORIUM CAPILLIFOLIUM (Lam.) Small IDC 96-5 (Chrysocoma? capillaris; in pascuis juxta Savanah). Type (Uttal 1984).

Aster concolor L.: *HAB.* in sylvis maritimis, a Virginia ad Floridam. ASTER CONCOLOR L. IDC 97-14 (Aster concolor; in sylvis maritimis a Virginia ad Floridam).

Aster carolinianus Walt.: *HAB.* in dumosis palustribus, a Carolina inferiore ad Floridam. ASTER CAROLINIANUS Walt. IDC 97-10 (Aster carolinianus; no locality given).

Aster cordifolius L.: *HAB.* a Canada ad Floridam, per tractus montium et in occidentalibus *Alleghanis*. ASTER DUMOSUS L. IDC 98-4 (Aster cordifolius; in aridis et apricis sylvarum).

Tussilago integrifolia Michx.: *HAB.* a Carolina ad Floridam, in sylvis humidis. (Perdicium *semiflosculare*. Walt.). CHAPTALIA TOMENTOSA Vent. IDC 99-20 (Tussilago integrifolia; no locality given). Type (Uttal 1984).

Doronicum nudicaule Michx.: *HAB.* in umbrosis sylvarum, a Virginia ad Floridam. ARNICA ACAULIS (Walt.) Britton et al. IDC 99-21 (Doronicum nudicaule; no locale given). Type (Uttal 1984). Species occurs today in a few counties of the Florida panhandle.

(***) Inula gossypina Michx.: *HAB.* in maritimis Carolinae et Floridae. CHRYSOPSIS GOSSYPINA (Michx.) Ell. IDC 100-3 (Inula gossypina; Lieux arides en Basse Caroline). Type (Uttal 1984).

(***) Inula graminifolia Michx.: *HAB.* a Carolina ad Floridam, frequens. (Erigeron *glandulosum*? Walt.). PITYOPSIS GRAMINIFOLIA (Michx.) Nutt. IDC 100-4 (Inula graminifol., Caroline). Type (Uttal 1984).

Erigeron canadense L.: *HAB.* a Canada ad Floridam. CONYZA CANADENSIS (L.) Cronq. IDC 100-7 (Erigeron canadense; Hab. in pascuis juxta Charleston in Carolina; two other localities are Malbaye and Illinois).

(***) Baccharis angustifolia Michx.: *HAB.* in scirpetis maritimis, a Carolina ad Floridam. BACCHARIS ANGUSTIFOLIA Michx. IDC 100-13 (Baccharis fol lineari integris frutex dioicus; B. linearifolius, no locality given. Another label gives angustifolia). Type (Uttal 1984).

Baccharis halimifolia L.: *HAB.* in maritimis, a Virginia ad Floridam. BACCHARIS HALIMIFOLIA L. IDC 100-11 (Baccharis halimifolia; no locality given).

(***) Conyza pycnostachya Michx.: *HAB.* a Carolina ad Floridam. (Gnaphalium *undulatum.* Walt.). PTEROCAULON PYCNOSTACHYUM (Michx.) Ell. IDC 100-17 (Conyza pycnostackia; no locality). IDC 100-20 (Conyza pycnostachya; Carolina). Type (Uttal 1984).

Buphthalmum frutescens L.: *HAB.* in maritimis, a Carolina ad Floridam. BORRICHIA FRUTESCENS (L.) DC. IDC 101-21 (Buphthalmum frutescens; Carolines).

Helianthus angustifolius L.: *HAB.* a Virginia ad Floridam. HELIANTHUS ANGUSTIFOLIUS L. IDC 104-6 (No. 9 Helianthus *angustifolius*; no locality given).

Galardia fimbriata Michx.: *HAB.* in paludosis apricis, a Carolina ad Floridam. HELENIUM DRUMMONDII H. Rock IDC 104-13 (Bellium grandiflorum. Lieux humides en Caroline et en Georgie, Galardia?). Type (Uttal 1984). Species does not occur in Florida today.

Galardia lanceolata Michx.: *HAB.* a Carolina ad Floridam, in aridis. GAILLARDIA AESTIVALIS (Walt.) H. Rock IDC 104-11 (Route de Sunbury 5 milles d Alatamha). Type (Uttal 1984).

Rudbeckia purpurea L.: *HAB.* a Virginia ad Floridam, in montosis. ECHINACEA PURPUREA (L.) Moench IDC 105-3 (No. 7 Rudb. purpurea; no locality given). Species occurs today in a few counties of the Florida panhandle.

Rudbeckia hirta L.: *HAB.* a Virginia ad Floridam. RUDBECKIA HIRTA L. IDC 104-15 (Rudbeckia hirta; de la haute Caroline).

(***) Silphium compositum Michx.: *HAB.* in sylvis maritimis, a Carolina ad Floridam (*S. laciniatum.* Walt.). SILPHIUM COMPOSITUM Michx. IDC 105-7 (Silphium compositum; in sylvis maritimis a Carolina ad Floridam). Type (Uttal 1984).

(***) Silphium pumilum Michx.: *HAB.* in Florida. BERLANDIERA PUMILA (Michx.) Nutt. IDC 105-18 (No 10 Silphium pumilum; in Georgia et Florida). Type (Uttal 1984).

Polymnia tetragonotheca L.: *HAB.* a Virginia ad Floridam. TETRAGONOTHECA HELIANTHOIDES L. IDC 106-1 (Polymnia tetragonotheca, no locality given).

Elephantopus scaber L.: *HAB.* a Virginia ad Floridam. ELEPHANTOPUS CAROLINIANUS Raeusch. IDC 106-11 (Elephantopus; no locality given).

CAMPANULACEAE

Lobelia crassiuscula Michx.: *HAB.* in paludosis herbosis Carolinae maritimae usque ad Floridam (*L. glandulosa*? Walt.). LOBELIA GLANDULOSA Walt. IDC 107-15 (Lobelia crassiuscula; in paludosis Carolina). Type (Uttal 1984).

ORCHIDACEAE

Satyrium repens L.: *HAB.* a Canada ad Floridam. GOODYERA REPENS var. OPHIOIDES Fernald IDC 108-2 (Satyrium repens? L.; Hudsonis ad Floridam in umbrosis sylvarum). Species does not occur in Florida today.

(***) Malaxis unifolia Michx.: *HAB.* in umbrosis sylvarum, a Carolina ad Floridam. MALAXIS UNIFOLIA Michx. IDC 108-3 (Malaxis unifolia; Swamps de la Basse Caroline). Type (Uttal 1984).

Limodorum tuberosum L.: *HAB.* a Canada ad Floridam. CALOPOGON TUBEROSUS (L.) Britton et al. IDC 108-9 (Limodorum? minus; Maryland & Virginie).

Arethusa divaricata L.: *HAB.* a Carolina ad Floridam, in maritimis. POGONIA DIVARICATA (L.) R. Br. IDC 108-14 (Arethusa divaricata; no locality given).

ARISTOLOCHIACEAE

Aristolochia serpentaria L.: *HAB.* a Pensylvania ad Floridam. ARISTOLOCHIA SERPENTARIA L. IDC 189-19 (Aristolochia; no locality given).

CANNACEAE

(*) Canna flava Michx.: CANNA FLACCIDA L. Species is not in the *Flora*, IDC collection, or the *Journal.* In a published paper on some rare and new plants of North America, written in 1792 by André Michaux and edited by Lamarck, Michaux cites this species as "Canna *Flava.* Cette plante habite les bords des rivières en Georgie et en Floride" (Rehder 1923).

ARACEAE

(**) Pistia spathulata Michx.: no locality given. PISTIA STRATIOTES L. IDC 109-3 (in rivulis ad Lacus George, Florida). In an archival box containing Michaux's cahiers at the Library of the American Philosophical Society in Philadelphia, there is a loose sheet titled, "Notes of topographic botany." Michaux wrote, "88 Pistia stratiotes dans les eaux de la Riv. St. Jean en Florida." Type (Uttal 1984).

PLATANACEAE

Platanus occidentalis L.: *HAB.* a lacu *Champlain* ad Ludoviciam et Floridam. PLATANUS OCCIDENTALIS L. IDC 109-5 (Platanus occidentalis; en remontant Misissipi, 3 li. au dessus de la Riv du grand Makaquita cad [c.a.d.] 37 lieues au dessus de la Prairie du Chien; Cet arbre abonde au Lac Champlain).

ERIOCAULACEAE

(**) Eriocaulon decangulare L.: *HAB.* in Pensylvania, Virginia et Carolina. (*E. serotinum*? Walt.). ERIOCAULON DECANGULARE L. IDC 109-10 (Eriocaulon decangulare; Hab. a Pensylvania ad Floridam). Seeds from Florida of this genus, presumably of this species, were sent to Thouin on January 5, 1789.

CYPERACEAE

(**) Scleria hirtella Sw.: *HAB.* in sylvis Carolinae. SCLERIA HIRTELLA Sw. IDC 109-17 (Scleria hirtella; Hab. in herbosis sylvaticus a Carolina ad Floridam).

Scleria interrupta Michx.: *HAB*. a Carolina ad Floridam, in pratis udis. (Obs. Eamdem in Gallo-Guyanna legit L. C. Richard.). SCLERIA HIRTELLA Sw. IDC 109-15 Scleria *interrupta*; Hab. in pratis udis a Carolina ad Floridam. Another label reads: (Scleria ciliata Juss.) Basse Carol. No. 76. Core (1936) indicates that *Scleria interrupta* Michx. is a synonym of *Scleria hirtella*.

(**) Carex leporina L.: *HAB*. a Canada ad Carolinam. CAREX SCOPARIA Schkuhr. IDC 110-21 (Carex leporina; Hab. a Canada ad Florida; In umbrosis Carolinae, Floridae).

URTICACEAE

Urtica cylindrica L.: *HAB*. in dumetosis, a Canada ad Floridam. BOEHMERIA CYLINDRICA (L.) Sw. IDC 111-11 (Urtica cylindrica; in Nova Caesarea).

MORACEAE

Morus rubra L.: *HAB*. a Canada ad Floridam. MORUS RUBRA L. IDC 111-14 (Morus rubra; in Maritimis Hab. in maritimis a Nova Caesarea ad Floridam et in Occidentalib. a Lacu Ontario).

(*) Ficus.: Floride. FICUS AUREA Nutt. IDC 106-10. Species not in the *Flora*; however, it is listed in the *Journal* and IDC collection.

ASTERACEAE

Ambrosia paniculata Michx.: *HAB*. a Canada ad Floridam. (Iva *monophylla*. Walt.). AMBROSIA ARTEMISIIFOLIA L. IDC 112-10 Ambrosia artemisifolia (no locality given). IDC 112-15 (Ambrosia paniculata elatior; no locality given). Type (Uttal 1984).

Iva frutescens L.: *HAB*. in maritimis, a Nova Anglia ad Floridam. IVA FRUTESCENS L. IDC 113-1 (Iva frutescens; no locality given. Flowers in August).

ARACEAE

(***) Calla sagittifolia Michx.: *HAB*. in paludosis Georgiae et Floridae. PELTANDRA SAGITTIFOLIA (Michx.) Morong IDC 113-7 (Calla sagittifolia; H. in paludosis Georgiae, Floridae). Type (Uttal 1984).

Arum dracontium L.: *HAB*. in umbrosis Carolinae et Floridae. ARISAEMA DRACONTIUM (L.) Schott IDC 113-11 (Arum dracontium; Hab. in Carolina, Florida).

JUGLANDACEAE

Juglans nigra L.: *HAB.* a Pensylvania ad Floridam. JUGLANS NIGRA L. IDC 114-5 (Juglans nigra; no locality given).
(*) Juglans hickory mentioned in the *Journal* is probably CARYA AQUATICA (Michx. f.) Nutt.

FAGACEAE

Castanea pumila Michx.: *HAB.* a Marilandia ad Floridam. CASTANEA PUMILA (L.) Mill. IDC 115-14 (Castanea pumila; a Virginia ad Floridam).
Quercus obtusiloba Michx.: *HAB.* a Canada et Nova Anglia ad Floridam. QUERCUS STELLATA Wang. IDC 117-15 (Pensylvania). Type (Uttal 1984).
Quercus alba L.: *HAB.* a Canada ad Floridam. (Var. *pinnatifida* Walt. *Car.* p. 235, n° 10.). QUERCUS ALBA L. IDC 118-2 (No. 12 Quercus alba virginiana Catesby; Basket oak, Quercus pinnatifida Walt; Chene blanc de Virginia: Charlenoix). IDC 118-4 (Quercus alba, in Carolinae).
Quercus prinus L.: *HAB.* in utraque Carolina, Georgia, Florida, &c. (Var. *palustris*). QUERCUS MICHAUXII Nutt. IDC 118-5 (No. 16 Q. prinus β palustris; no locality given). IDC 118-11 (No. 13 Q. prinus β palustris, Illinois).
Quercus virens Ait.: *HAB.* a Virginia ad Floridam. QUERCUS VIRGINIANA Mill. IDC 118-12 (Quercus virens Ait.; Hab. a Virginia maritima ad Floridam).
Quercus phellos L.: *HAB.* a New-Jersey ad Floridam. (Var. *sylvatica*). QUERCUS PHELLOS L. IDC 118-20 (Quercus phellos; Hab. a Pensylvania ad Floridam). *Journal* plant.
Quercus aquatica Catesb.: *HAB.* a Marilandia ad Floridam. QUERCUS NIGRA L. IDC 116-1 (Quercus nigra aquatica [nigra crossed out], No. 5; no locality given).
Quercus nigra Catesb.: *HAB.* a Marilandia ad Floridam. QUERCUS NIGRA L. IDC 116-3 (No. 8 Quercus nigra; Sur les bords de la Mer en Floride).
(***) Quercus falcata Michx.: *HAB.* a Virginia ad Floridam. QUERCUS FALCATA Michx. IDC 116-12 (Q. falcata; haute et Basse Caroline). Type (Uttal 1984).

BETULACEAE

Corylus americana Walt.: *HAB.* a Canada ad Floridam, etiam in maritimis. CORYLUS AMERICANA Walt. IDC 118-21 (Environs de New York, Montagne Pensylvania). Species does not occur in Florida today.

Carpinus americana Michx.: *HAB.* a Canada ad Floridam. CASTANEA DENTATA (Marshall) Borkh. IDC 115-12. (Fagus *Castanea*: *β americana;* Castanea vesca americana). Species occurs in extreme western part of the Florida panhandle.

HAMAMELIDACEAE

Liquidambar styraciflua L.: *HAB.* a Pensylvania ad Floridam. LIQUIDAMBAR STYRACIFLUA L. Species neither in the IDC nor in the *Journal.*

PINACEAE

Pinus palustris Mill.: *HAB.* a Carolina septentrionali ad Floridam, praesertim in maritimis. PINUS PALUSTRIS Mill. IDC 119-5 (Specimens given are unlabeled cones, but species name is written in list form with other *Pinus* species).

(**) Pinus taeda L.: *HAB.* in Virginia, Carolina et Georgia. PINUS TAEDA L. IDC 119-5 (Specimens given are unlabeled cones, but species name is written in list form with other *Pinus* species). *Journal* plant.

(*) Pinus follis binis. This is probably PINUS CLAUSA (Chapm. ex Engelm.) Vasey ex Sarg. Not in the *Flora; Journal* plant.

CUPRESSACEAE

Cupressus disticha L.: *HAB.* a Marylandia ad Floridam et a Tennassee fluvio ad Ludoviciam. TAXODIUM DISTICHUM (L.) L.C. Rich. IDC 119-10 (Cupressus disticha; Le 3 Mars 1794 pres Wilmington. Another label reads Carol & Georg). *Journal* plant.

EUPHORBIACEAE

(**) Phyllanthus carolinensis Walt.: *HAB.* in Carolina et Georgia. PHYLLANTHUS CAROLINENSIS Walt. IDC 119-13 (Phyllanthus in Maritimis Georgiae, Floridae).

Euphorbia graminifolia Michx.: *HAB.* in maritimis Floridae et Georgiae. (*Obs.* Affinis *E. hissopifoliae.*). POINSETTIA CYATHOPHORA (Murray) Bartl. IDC 119-19 (Euphorbia graminifolia; Partis maritimes et les plus meridionales de la Georgie). Type (Uttal 1984).

Euphorbia heterophylla L.: *HAB.* in Florida. POINSETTIA HETER-

OPHYLLA (L) Klotzsch & Garcke ex Klotzsch IDC 119-18 (Floride; famille des Euphorbes?).

Euphorbia maculata L.: *HAB.* in Florida. CHAMAESYCE MACULATA (L.) Small IDC 120-2 (Euphorbia maculata; Floride).

Stillingia sylvatica L.: *HAB.* a Carolina ad Floridam. STILLINGIA SYLVATICA L. IDC 119-15 (Stillingia sylvatica; Basse Carol. et Georg.). *Journal* plant.

(***) Croton argyranthemum Michx.: *HAB.* in sylvis aridis Georgiae et Floridae. CROTON ARGYRANTHEMUS Michx. IDC 120-12 (Croton argyranthemum; Floridae). Type (Uttal 1984).

(*) Hippomane. HIPPOMANE. Species neither in the *Flora* nor in the *Journal.* Genus listed in shipments of seeds from Florida sent to d'Angiviller on August 2, 1788, and to Thouin on January 5, 1789. In the shipment to Thouin, Michaux wrote Hippomane herbacée, which indicates that the seeds are probably not those of the manchineel, *Hippomane mancinella* L. Furthermore, the manchineel does not range in the area Michaux traveled.

Acalypha caroliniana Walt.: *HAB.* a Virginia ad Floridam. ACALYPHA OSTRYIFOLIA Riddell IDC 120-14 (Acalypha caroliniana; no locality given).

(***) Jatropha stimulosa Michx.: *HAB.* a Virginia ad Floridam. (*J. urens.* Walt.). CNIDOSCOLUS STIMULOSUS (Michx.) Engelm. & A. Gray IDC 120-15 (Jatropha stimulosa herbacea; no locality given).

CUCURBITACEAE

(*) Wild squash. This is probably the Okeechobee gourd, CUCURBITA OKEECHOBEENSIS (Small) L. H. Bailey. Both John and William Bartram found this species on the St. Johns River in the vicinity where Michaux located his gourd. Not in the *Flora*; *Journal* plant.

HYDROCHARITACEAE

(***) Vallisneria americana Michx.: *HAB.* in flumine *Mississipi* et in fluvio *S. Joannis* Floridae. VALLISNERIA AMERICANA Michx. IDC 120-20 (plant unidentified on IDC; no locality given). Type (Uttal 1984).

EMPETRACEAE

(***) Ceratiola ericoides Michx.: *HAB.* in aridis sabulosis Georgiae et Floridae. CERATIOLA ERICOIDES Michx. IDC 120-21 (Ceratiola ericoides; Hab in Georgia, Florida). Type (Uttal 1984).

OLEACEAE

Olea americana Michx.: *HAB.* in maritimis, a Carolina ad Floridam. OSMANTHUS AMERICANUS (L.) Benth. & Hook. f. ex A. Gray IDC 121-1 (Olea americana; no locality given). *Journal* plant.

Adelia porulosa Michx.: *HAB.* in maritimis Floridae. FORESTIERA SEGREGATA (Jacq.) Krug & Urban IDC 121-4 (Adelia porulosa; Hab in Maritimis Floridae). Type (Uttal 1984).

MYRICACEAE

Myrica cerifera L.: *HAB.* a Carolina ad Floridam, in umbrosis. (Var. *arborescens*); a Nova Anglia ad Floridam, in udis et juxta rivulos. (Var. *media*); in aridis, a Carolina ad Floridam. (Var. *pumila*). MYRICA CERIFERA L. IDC 121-21 (Myrica cerifera arborescens; no locality given). IDC 122-1 (Myrica cerifera pumila; no locality given). IDC 122-3 (Myrica cerifera media; no locality given). *Journal* plant.

AQUIFOLIACEAE

Ilex opaca Ait.: *HAB.* in umbrosis humidis, a Pensylvania ad Floridam. ILEX OPACA Ait. IDC 122-11 (one label gives Ilex opaca, Florida; another label gives Ilex opaca, Carol).

Ilex dahoon Walt.: *HAB.* in paludosis, a Carolina ad Floridam. (*I. Cassine.* Linn. Ait.) ILEX CASSINE L. IDC 122-8 (Ilex cassine de 50 pieds de haut aux lieux humides, Floride).

Ilex cassena Michx.: *HAB.* in maritimis, a Carolina ad Floridam (*I. Cassine.* Walt.; Cassena *vera.* Catesb.). ILEX VOMITORIA Ait. IDC 122-7 (Floride; Ilex *cassena* vomitora Ait.; Cassena vera Catesby. Two other labels read Ilex cassena vomitoria Ait. and Ilex cassena vera, Yapon). *Journal* plant.

Prinos glaber L.: *HAB.* a Nova Anglia ad Floridam. (*Obs.* An vere congener *P. verticillati.* Linnaei?). ILEX GLABRA (L.) A. Gray IDC 123-21 (Prinos glaber; de Florida).

VITACEAE

Vitis labrusca L.: Hab. a Pensylvania ad Floridam. (*V. taurina.* Walt.). VITIS LABRUSCA L. IDC 132-6 (Vitis labrusca; Fox grape, in Pensylvania, Carolina). Not found in Florida today.

Vitis cordifolia Michx.: *HAB.* a Pensylvania ad Floridam. VITIS CINEREA (Engelm.) Engelm. ex Millardet IDC 123-3 (Vitis cordifolia Winter grape; Hab. a Nova Anglia ad Carolinam). IDC 123-4 (No. 3; no locality given). Type (Uttal 1984). Species occurs today in the western part of Florida and the panhandle.

(***) Vitis rotundifolia Michx.: *HAB.* a Virginia ad Floridam. (Obs. *Muscadin grape* vulgo audit.). VITIS ROTUNDIFOLIA Michx. IDC 122-21 (Vitis rotundifolia; no locality given). Type (Uttal 1984).

AMARANTHACEAE

(*) Annual 12 ft tall. This plant, described in the *Journal*, is probably AMARANTHUS AUSTRALIS (A. Gray) Sauer. The species is common today in wet disturbed sites where Michaux traveled along Florida's east coast. Not in the *Flora*. Tab. 50 of the *Flora*, which illustrates *Acnida rusocarpa* Michx., is probably *Amaranthus australis*. IDC 123-12 (Amaranthus altissimus, Acnida) appears to be this taxon.

RUTACEAE

Zanthoxylum tricarpum Michx.: *HAB.* a Carolina ad Floridam. (*Z. fraxinifolium*? Walt.). ZANTHOXYLUM CLAVA-HERCULIS L. IDC 123- 15 (Zanthoxylum tricarpum; no locality given). Type (Uttal 1984).

SMILACACEAE

Smilax bona nox L.: *HAB.* in maritimis Carolinae et Floridae. SMILAX BONA-NOX L. IDC 124-7 (Smilax bona nox; no locality given).

(**) Smilax laurifolia L.: *HAB.* in Carolina et Georgia. SMILAX LAURIFOLIA L. IDC 124-9 (Smilax laurifolia; Basse Carol. et Floride).

MENISPERMACEAE

Menispermum carolinianum L.: *HAB.* in Carolina, Georgia, Florida. COCCULUS CAROLINUS (L.) DC. IDC 124-20 (Menispermum carolinum; no locality given).

ZAMIACEAE

Zamia pumila Michx.: *HAB.* in Florida. ZAMIA PUMILA L. IDC 125-2 (Zamia pumila; in Florida). Species listed in a shipment of seeds from Florida sent to Thouin on January 5, 1789. *Journal* plant.

CUPRESSACEAE

Juniperus virginiana Michx.: *HAB.* a lacu Champlain ad Floridam. JUNIPERUS VIRGINIANA L. IDC 126-7 (Juniperus virginiana; Hab. a Lacu Champlain ad Floridam).

Juniperus rbadensis [barbadensis] L.: *HAB.* in Floridae maritimis et in *Bahama.* JUNIPERUS SILICICOLA (SMALL) BAILEY. IDC 126-8 (Juniperus barbadensis; Hab. in Floridae maritimis, Bahamae).

ACERACEAE

Acer montanum Ait.: *HAB.* in Canada, raro in montibus excelsis *Alléghanis*, usque ad Floridam. ACER SPICATUM Lam. IDC 127-6 (Acer Pensylvanium *montanum* [Pensylvanium crossed out]; Lacs Mistassins Canade & tres hautes Montagnes de l'Ameriq. Septentrionale). Species does not occur today in Florida.

(**) Acer rubrum L.: *HAB.* a Pensylvania ad Carolinam. ACER RUBRUM L. IDC 127-8 (Acer rubrum; Carolina). *Journal* plant.

FABACEAE

Mimosa horridula Michx.: *HAB.* a Virginia ad Floridam (*M. Intsia.* Walt.). *Obs.* Affinis *M. quadrivalvi.* MIMOSA QUADRIVALVIS L. IDC 127-11 & IDC 127-12 are Mimosa quadrivalvis. IDC 127-13 (Mimosa vilotica?; Environs de St Augustin; another label gives Mimosa herbacea, Illinois; a third label gives Mimosa herbacea, du Missouri). Type (Uttal 1984).

Gleditsia monosperma Walt.: *HAB.* in Carolina, Georgia, Florida et regione Illinoensi. GLEDITSIA AQUATICA Marshall IDC 128-10 (Gleditsia monosperma; no locality given). *Journal* plant.

NYSSACEAE

Nyssa tomentosa Michx.: *HAB.* ad amnem *S. Mary* et in Florida. NYSSA AQUATICA L. IDC 129-7 (Nyssa tomentosa, Riv. Altamaha, S[te] Mary et Floridae). IDC 129-8 (Nyssa tomentosa; Hab. ad amnes S[t] Mary, Altamaha, et in Floridae). Type (Uttal 1984).

VITTARIACEAE

Vittaria angustifrons Michx.: *HAB.* in Florida, juxta amnem *Aisa-hatcha.* VITTARIA LINEATA (L.) Smith IDC 129-10 (Pteris lineata, Sur les Bords de la Riv Aisahatcha [Indian River, Brevard Co., Fla.] le 1[er] Avril, Floride). *Journal* plant. Type (Uttal 1984).

DENNSTAEDTIACEAE

Pteris aquilina L.: *HAB.* a sinu *Hudson* ad Floridam. PTERIDIUM AQUILINUM (L.) Kuhn. IDC 129-11 (Pteris aquilina; Canada).

BLECHNACEAE

Blechnum serrulatum Michx.: *HAB.* in Florida, juxta amnem *Aisa-hatcha.* BLECHNUM SERRULATUM L. C. Rich. There is no specimen of this species in the IDC collection; however, according to Morton (1967), a specimen collected by Michaux in Florida exists in the Jussieu Herbarium (Cat. no. 1388A). *Journal* plant. Type (Uttal 1984).

POLYPODIACEAE

(**) Polypodium vulgare L.: *HAB.* a Canada ad Carolinam. (*P. virginianum*? Linn.). POLYPODIUM VIRGINIANUM L. IDC 130-12 (Polypodium vulgare; Hab. in arborib. a Canada ad Floridam). Species does not occur today in Florida.

Polypodium ceteraccinum Michx.: *HAB.* parasiticum in Kentucky, Tennassee, Florida. (Achrostichum *polypodioides* Linn.). PLEOPELTIS POLYPODIOIDES (L.) E. G. Andrews & Windham IDC 130-19 (in arboribus Floride, Achrostichum polypodioides). In the Jussieu Herbarium (Cat. 1090A) there is a specimen of *Polypodium polypodioides* collected in Florida by André Michaux (Morton 1967).

Polypodium sp. indet. Florida, collected by Michaux. Morton (1967) reports that this plant was identified as *Polypodium plumula* Humb. & Bonpl. ex Willd. PECLUMA PLUMULA (Humb. & Bonpl. ex Willd.) M. G. Price. Morton remarks that he did not study the specimen critically, but that the fern was *Polypodium plumula* Humb. & Bonpl. ex Willd., *Polypodium pectinatum* L., or an allied taxon. There is a *Polypodium plumula* Willd. determined by D. C. Eaton in the Paris herbarium (IDC 130-3) that lacks a locality.

PTERIDACEAE

Acrostichum aureum L.: *HAB.* in Florida, ad amnem *Aisa-hatcha.* ACROSTICHUM DANAEIFOLIUM Langsd. & Fisch. IDC 130-18 (Acrostichum aureum; sur la rivi[ère] Aisa hatcha Floride). Morton (1967) states that the specimen in the Paris herbarium is *A. danaeifolium* and not *A. aureum.*

OSMUNDACEAE

Osmunda regalis L.: *HAB.* a Canada ad Floridam. OSMUNDA REGALIS L. IDC 131-2. (Osmunda regalis; Hab. a Nova Anglia ad Carolinam). IDC 131-3 (Osmunda regalis; in Canada).

PSILOTACEAE

Psilotum floridanum Michx.: *HAB.* in Florida. PSILOTUM NUDUM (L.) P. Beauv. IDC 132-12 . (No 45 2 Miles du Fort Matança, Floride, sous les Orangers. Another label reads Lycopod nudum Floride). Type (Morton 1967).

SELAGINELLACEAE

Lycopodium apodum L.: *HAB.* in umbrosis sylvarum herbosis, a Virginia ad Floridam. SELAGINELLA APODA (L.) Fernald IDC 132-15 (Lycopodium apodum; a Pensylvania ad Floridam; another label reads de Caroline et de Georgie, Parties Maritimes).

Appendix 2

Letters and Documents Pertaining to André Michaux's Mission in America

2-1. Baptismal record of André Michaux. Archives Municipales de Versailles GG 345.
2-2. Letter from l'Abbé Nolin to Madame Helvetius, April 18, 1784.
2-3. Letter from Count de Vergennes, minister of foreign affairs, to Benjamin Franklin, August 9, 1785, Paris.
2-4. Letter from André Michaux to Cuvillier, July 15, 1786, New York.
2-5. Letter from André Michaux to Count d'Angiviller, July 15, 1786, New York.
2-6. Letter from André Michaux to Count d'Angiviller, December 26, 1786, Charleston.
2-7. Letter from André Michaux to Count d'Angiviller, April 8, 1787, Charleston.
2-8. Letter from André Michaux to André Thouin, November 6, 1787, Charleston.
2-9. Letter from André Michaux to Count d'Angiviller, February 9, 1788, Charleston.
2-10. Letter from André Michaux to Count d'Angiviller, April 24, 1788, St. Augustine.
2-11. Letter from André Michaux to Count d'Angiviller, June 10, 1788, Charleston.
2-12. Letter from André Michaux to Count d'Angiviller, July 1, 1788, Charleston.
2-13. Letter from André Michaux to Count d'Angiviller, August 2, 1788, Charleston.
2-14. Letter from André Michaux to Count d'Angiviller, October 1, 1788, Charleston.

2-15. Letter from André Michaux to Count d'Angiviller, January 5, 1789, Charleston.

2-16. Letter from André Michaux to Monsieur, brother of Louis XVI, addressed to Count d'Angiviller, January 5, 1789, Charleston.

2-17. Letter from André Michaux to Count d'Angiviller, April 28, 1789, Charleston.

2-18. Letter written by André Michaux to Governor Vizente Manuel Zéspedes requesting permission to explore Spanish East Florida. See Photo 12.

André Michaux

Lan mil sept cent quarante six le huitieme jour
du mois de Mars andré né ce meme jour fils de
andré Michaux fermier du Roy a Satory et de
Marie charlotte barbet son epouse a été baptisé
par nous soussigné pretre de la mission faisant
les fonctions curialles, le parain a été claude
barbet oncle de lenfant, et la maraine Marie
barbe Michaux tante de lenfant lesquels ont
signés avec nous, le pere absent. Claude Barbée
marie Barbe michaux Armand Prêtre

2-1. Baptismal record of André Michaux. Archives Municipales de Versailles GG 345.

2-2. Unpublished letter from l'Abbé Nolin to Madame Helvetius. From Yale University Library, The Franklin Papers. Permission from the American Philosophical Sociey, Philadelphia.

Paris, 18 April 1784

Madam

Mr. Franklin has translated and has had published a Catalogue of trees and shrubs which Mr. Bertrand [Bartram] is selling in the neighborhood of Philadelphia. I would like to have a look at the original in English. If Mr. Franklin is willing to lend it to you, I will return it very soon. I am in a hurry to have it.

I have the honor to be with respect, dear Madam, your humble and very obedient servant.

L'Abbé Nolin

2-3. Unpublished letter from Charles Gravier, Comte de Vergennes (1719-1787), minister of foreign affairs under Louis XVI, to Benjamin Franklin. Original in the Département des Archives Historiques in the series, *Correspondance politique*, subseries *United States*, volume *30*, fol. 189 (r°-v°). From Yale University Library, the Franklin Papers. Permission from the Archives du Ministère des Affaires Etrangères.

[Paris] 9 August 1785

This letter, Sir, will be presented to you by Mr. Michau [Michaux] botanist, associated with the King's nurseries, under the order of Monsieur le Comte d'Angiviller, Director of His Majesty's [properties]. Mr. Michau [Michaux] having been well regarded by his work and his discoveries on the culture [of plants] on his trip of several years in Asia, the King has requested that he continue his research in other climates starting with North America; he has received the order to go there and establish a collection of trees, shrubs, plants, and seeds from the New World, little known or unknown in France and of which His Majesty wants to introduce the cultivation on His Properties. Our botanist having received his instruction from the Director general of Properties, is getting ready to leave on the new trip and I hope that he will find that you have arrived in good health in your Country. I expect Sir, from your love of science, because of your affection for France, who misses you, and for your friendship towards me, that you will under these circumstances help in the King's wishes and in the success of Mr. Michau's [Michaux's] mission, both by your wise counsel as well as by that of bright agriculturists with whom he must be in contact. I pray you to help him in ways that will depend on you, independently of the assistance that will be forthcoming from Monsieur Otto, to whom I have likewise written.

Receive, Sir, the new assurances of my wishes for the general prosperity of the United States and for your happiness in particular.

I have the honor to be very certainly Mr. Franklin . . . .

To Mr. Franklin

2-4. This letter is probably to Monsieur Cuvillier, personal secretary of Count d'Angiviller. From a copy in the Henry Savage research collection, South Caroliniana Library, University of South Carolina. Permission from the Bibliothèque Centrale du Muséum National d'Histoire Naturelle, Paris.

Monsieur

I have just announced to Monsieur Le Comte that I have drawn the sum of 5000# from Monsieur Dutartre. I have not made many extraordinary expenditures since the letter I had the honor to write you in the month of May last, but more considerable expenses are constantly presenting themselves; however, the sum will suffice probably up to at least September and October. Expenses will increase only in proportion to the quantity of shipments made. Henceforth I will be always ready to produce the account of my expenses. I have collected together some papers scattered due to the changing of dwelling places to which I was often exposed in the first month after my arrival here.

Last month I visited Pennsylvania, Maryland, and some parts of Virginia. Philadelphia deserves to be compared with the best cities of Europe after Paris. At first sight it is very pleasant because of the regularity of its streets and for its situation, merchandise from all parts of Europe are plentiful. There are more well-informed people and scholars than in any other parts of this continent. Foreigners complain of the few entertainments one finds there in the societies, because the Quakers who form the greatest part of the inhabitants are very religious, and there are no theaters in this city. One judges the opulence of the inhabitants of the province by the wealth, good quality, and number of their equipage, horses and carriages, which bring provisions to the city. In the country the good quality of the land which is clayey and substantial, combined with a warm and humid climate, results in great fertility. The inhabitants of this province, German in origin, well deserve these advantages; there is not a more industrious people among the different Nations which have come to form the United States. Maryland is perhaps the most sterile of all the United States. The soil is sandy and rye grows there with difficulty. The portions of Virginia which I have visited are good and are likely to be fertile, but inhabited by large landowners with great estates. They are not very active or industrious and the land is cultivated only in proportion to the number of slaves.

I am with all affection and regards

Monsieur

Your very humble and very obedient Servant

A. Michaux

New York, this 15 July 1786

2-5. From a copy in the Henry Savage research collection, South Caroliniana Library, University of South Carolina. Permission from the Bibliothèque Centrale du Muséum National d'Histoire Naturelle, Paris.

Monsieur Le Comte
I have had the honor of informing you in my letter of 18 January that I have not made use of my letter of credit on Mr. Morris because his manner of evaluating the Piastre [dollar] at 108 S. is a real loss for the Department. Monsieur La Forest, Consul of Finance here was able to show me the way to have Piastres at a better rate. I send back to you then this letter for 1500 # and I inform you that I have drawn on Monsieur Dutartre for the sum of 10,000 # at 103 S per Piastre.

I have the honor of informing you today, Monsieur Le Comte, that I have just drawn on Monsieur Dutartre the sum of 5000 # remaining from that which I was empowered to receive from Mr. Morris. With the help of Monsieur La Forest I have had the Piastres at 102S6 which made the Department a profit of 120 #.. 7S6 at the present rate of 105S per Piastre at the chamber of commerce in New York.

I announced to you Monsieur Le Comte, my departure with date for Philadelphia. I visited Pennsylvania, Maryland, and a part of Virginia as far as Fredericktown and Shennadoa River; I gathered a great many interesting plants, but I did not remain because the essential purpose is to gather seeds, and because I would have made expenditures for which one could not extract the full advantage, which indisputably can be hoped for in two months. I have been very well received by Docteur Franklin and all the persons to whom I have had introductions. Docteur Franklin has promised me letters of Recommendation for Carolina and Georgia. General Washington has offered to allow me to send my collections to his home on deposit and I will profit on all the occasions by the offers which have been made to me.

I shall set out before the 15 of next month for Carolina and Georgia, so that by arriving soon enough for seeds, I can reap every advantage possible from the journey.

I am with very profound Respect
Monsieur Le Comte,
New York, the 15 July 1786 Your very humble and obedient servant
A. Michaux

2-6. From a copy in the Henry Savage research collection, South Caroliniana Library, University of South Carolina. Permission from the Bibliothèque Centrale du Muséum National d'Histoire Naturelle, Paris.

Monsieur Le Comte

The scarcity of boats destined for New York has prevented me from making any shipments to you until this moment. I profited from an American packet-boat to send you thirteen boxes of trees and three of seeds. If you agree, there is one of these boxes destined for the Botanic Garden in Paris and in the boxes of seeds a little package to the address of Monsieur Thouin. You will find attached the lists of the trees and that of the seeds. I am a little fearful about their arrival in New York, for there are several trees susceptible to freezing in the boxes if the cold there is too severe. To prevent this danger, I am ordering the gardener to go to the disembarkment and to transport the boxes immediately on board the French packet-boat if it is there, because the plants do not freeze on the Ships. It is because of these difficulties that I would desire for next year a French packet-boat to Charleston. I have been promised that there will be in a month a Ship destined for L'Orient. I will send you then a good shipment and the Plants will not be exposed to the dangers of the Freeze. I would like to send you in this way some wild turkeys but I do not dare to hope for this. One can take them living only in February and even then they die if one does not give them a great deal of space.

The best way is to incubate the eggs which are found rather easily, it is the same with the Canards d'Eté (summer ducks). I will do everything possible in this matter and I do not doubt at all of success. I have already finished the fences in the place of cultivation which I announced to you, Monsieur Le Comte. I have worked unceasingly and in spite of the expenses of a new establishment which consist in repairs, purchases of work tools, living expenses for men and horses, which the following years will be taking out of the products of the land, it will be nevertheless a great deal more economical than if I had remained in the city. It is true, Monsieur Le Comte that in spite of the solitude and the pain that I have given myself the quantity of my collections will not equal this year those which I can make the following years, but if I had remained in the city they would have been less and more expensive.

The excess labor to which I have been subjected since my arrival without being assisted except only rather weakly by my son, had obliged me to speak poorly of him to you. He has been very aware of it and from that time on the firmness and strictness which I have been obliged to use have produced a good effect. Thoughtfulness aside, I am content with him for he seems to do all that he can and to be interested in our affairs and this is what I principally desired.

In addition to the seeds of this shipment there are several others which I could now send you but for the difficulty of drying them; you will receive them, Monsieur Le Comte, only at the end of Spring. There is a kind of white doe so rare in this Country that it has not been seen for many years. Someone has captured two females and has promised me one of them. If it is as interesting as the description that I have been given, I will devote all of my attention to sending the species to France. The work with which I am entrusted is very interesting but there is a great deal to do, for every day I recognize something new, which will not be neglected because of my taking note of everything so as to profit by it when circumstances permit. The buffaloes of America differ a great deal from those of the old continent. They will be useful for their skin and their flesh which is said excellent to eat. I know that the skin of those of the [Near] East is more useful than that of all the other animals but the flesh is bad and in Bagdad only the most unfortunate Arabs nourished themselves on it.

I have already made plans for the placement of different acquisitions in the garden and I feel sure that you will be pleased with my work. I shall send you, Monsieur Le Comte, ten of the same partridges which you received last year but I will wait until I have some bred in captivity whose eggs I will have brooded by hens. The departure of a ship for L'Orient has just been confirmed by the announcement that the ship owner has made of it in the public papers, and I am going to work on a shipment and make it as interesting as possible, but with regard to the trees, shrubs, and other plants one will enjoy them with much more satisfaction if placed in a kind of great orangerie in which all the trees would be in the soil itself. It would not even be necessary to make any fire, because the trees would not be damaged by a little frost which might penetrate, if one took the precaution of not transplanting those which one would judge appropriate, at the approach of the great cold spells. I have experienced that here where the freezes are very weak, Magnolia grandiflora, Gordonia and several other trees have frozen leaves, Bignonia sempervirens has been so much frozen that the bark has been entirely stripped away from them. One could even make two orangeries, one ornamental and the other a Nursery less expensive in terms of construction.

I have just received news from New York. I had left, when setting out, about two hundred Piastres for the service of the establishment. The gardener informs me that Monsieur de la Forest makes difficulty in advancing him some money beyond that sum which is used up. This obliges me to send via New York a bill of exchange for 1000 pounds on Monsieur Dutartre payable to the order of Monsieur de la Forest, but I am ordering the gardener not to ask for money in the future without sending me the report of his preceding expenditures, and I myself will not let the month of January pass without sending the expenses for the preceding year.

I am with a very profound Respect
Monsieur Le Comte
Your very humble and very obedient servant
A Michaux

Charleston the 26 December 1786

I request permission to insert here a letter for my brother in which I request some tools for the service of the Department here.

List of Trees, Shrubs, and Plants from the Carolinas, 26 December 1786

First Case

1300 Laurus borbonia. This tree which is evergreen has pubescent leaves underneath. It is found commonly in very wet areas but in drier places the leaves are glabrous and shining.

50 Laurus aestivalis. This shrub grows in areas that are always submerged. The roots are always large, bent and it is difficult to dig it up. One must go in the water up to one's knees.

60 Laurus ? new species. This shrub is very fragrant. I have never seen any larger than these. It grows in submerged areas. The flowers are sessile while those of the preceding species are pedicellate.

3 Laurus benzoin. It grows in shady moist areas. Spice-bush.

2nd Case

300 Laurus sassafras. This tree grows in sandy and dry areas.

3rd Case

20 Ilex aquifolium americanum. This shrub becomes very large. It differs little in its leaves from the European one but the fruits are bigger. It grows best in moist areas.

20 Ilex caroliniana. This shrub grows in very moist areas.

350 Ilex cassine. This shrub does not lose its leaves in the winter, but it bears so many red fruits that one is not aware of leaves at a short distance. It is named here Yapon and sometimes cassine. It grows in sandy areas and spreads horizontally.

10 Ilex with deciduous leaves. I think this is a new species. It is really in the genus Ilex by its fruits, but it loses its leaves.

25 shrubs of the family of Ilex, but the calyx has 5 parts and the seeds are in 5. The fruit is larger than those of others, it loses its leaves very early. It grows in sandy and cool areas.

4th Case

200 Magnolia grandiflora. There are some trees in the forest that are three and one-half ft in diameter. It grows where the soil has been enriched by the decay of other vegetation, but nevertheless sandy.

40 Magnolia tripetala. The odor of this shrub is very good; I have not seen any larger than a fist, but nevertheless it is at higher elevation than other shrubs of this size and this makes me believe that I have not yet seen it in the ideal situation and that it was brought by the birds.

10 Magnolia glauca. This tree grows here in moist and sometimes aquatic areas and it is more than one-half foot in diameter. All moist areas are covered by this plant, but it is difficult to find young plants.

12 Sarracenia lutea and purpurea. These plants especially the purple one grow in very moist areas.

5 Pinus foliis longissimus. This tree grows in sandy, moist and warm areas.

1 Lianne Supple-Jack. This climbing shrub has not yet been described by botanists and I give its common name. In opening the box it will be necessary to hold the plant with two hands after having removed the moss to inhibit its possible damage because of its elasticity.

5th Case

30 Nyssa montana. This tree differs by the larger size of its fruit from that of the one from New Jersey. It grows in very damp places.
2 Nyssa aquatica. This tree grows in permanently submerged areas. It differs much from the preceding species by the shape of its seeds and other characteristics. It is a very tall tree.
36 Cupressus disticha. This tree is very tall and stout; it grows in areas that are permanently submerged.
150 Gordonia lasianthus. Large tree, evergreen. It grows in the form of a sugar pine and is only found in damp places and submerged about nine months of the year—does not do well in dry places.
12 Viburnum cassinoides and Viburnum nudum? V. cassinoides grows in moist areas and appears not to like frost having been damaged after I dug it up. V. nudum? whose leaves are very large grows in aquatic areas.

6th Case

2 Styrax occidentalis. Rather moist places.
2 Hopea tinctoria. Sandy and moist areas but I have not seen enough to know what soil is best.
3 Stewartia malacodendron. Low and humid areas, the monstrous and twisted roots kept me from sending many of these trees and many others. Next year I will be in these southern areas in time to observe and recognize them as now the trees are leafless and are difficult to recognize.
3 Sideroxilon lycoides? Large shrub that grows in the woods near streams. There is another with silvery leaves that grows in dry sandy soil but I only found it as a small shrub.
12 Unknown tree Daphne? This shrub grows only in poor soils. It is five inches in diameter, it sheds its leaves only as new ones are coming in.
4 Diospiros. There are two varieties of this tree whose fruit is very good to eat, but it should be cultivated as an espalier. The difficulty in finding it with deep roots has kept me from sending more, as it is very common here.
50 Zanthoxilon. This shrub grows to the height of approximately 30 feet and approximately 6 inches in diameter. It is found in sandy dry areas.

100 Magnolia grandiflora.
30 Sarracenia lutea.

7th Case

24 Juglans hickory.
3 Liriodendron.
12 Liquidambar. This tree grows in moist areas. The wood is white, in New Jersey there is a species with red wood of which furniture is made.
6 Castanea pumila; the chinquapin likes sandy areas.
35 Oaks of several species.
300 Oaks with willow leaves.
15 Sarracenia.

8th Case

75 Andromeda mariana. This shrub grows in moist places. It is two feet high.
30 Andromeda sempervirens. This shrub is five feet high. It is evergreen and grows in very wet areas.
25 Andromeda racemosa. Humid sites.
10 Andromeda paniculata. Humid sites. The flowers of this one are greenish in color and not very pretty.
10 Cyrilla racemiflora. This small tree is about 25 feet in height. It grows in aquatic areas and even submerged. One can find stems of 4-5 inches in diameter. I think it is the Andromeda plumata of Bartram.

9th Case

Andromeda sempervirens. See comments above.
25 Prunus americanus.

10th Case

25 Azalea. Two species of humid areas.
100 Vaccinium. Two species.
100 Fothergilla. Very moist areas.
—Hypericum and Ascyrum. Sandy and moist areas.
50 Clethra. Moist areas.
12 Myrica cerifera humilis. Dry areas but sometimes moist.
10 Myrica cerifera arborea. Very moist areas.

11th Case

100 Bignonia sempervirens. This small climbing tree is evergreen. It spreads on the shrubbery and on trees of medium height. In old forests one sees specimens with a diameter of one inch. It likes moist areas. In the winter it must be placed in an orangerie. All those that I brought for shipping were frozen in one night where the temperature went down to 5 degrees.
10 Bignonia crucigera. This shrub climbs on top of the tallest trees and obtains its substance in the manner of European ivy and of Toxicodendron. It grows in the oldest forests in aquatic areas. It does not seem to be affected by the cold.
10 Liannes-Supple-Jack. Very moist areas and even submerged nine months.
12 Glycine frutescens. Very humid areas and rivers with flowing water.
12 Smilax laurifolia. This species covers the tallest trees, it keeps its leaves during the winter. It is found in aquatic and submerged areas.
30 Cissampelos? Species of Smilax with red fruit and one seed. Areas dry and sandy.
1 Erythrina herbacea. Long lived plant whose seed is of the leguminous type, red like coral. The roots are up to nine inches in diameter. Sandy soils.

12th Case

Magnolia grandiflora.
Magnolia tripetala.
Gordonia lasianthus.
Pinus foliis longissimus.
Chamaerops (caroliniana) palmeto of Carolina.

List of Seeds of Carolina 26 December 1786

Gordonia lasianthus.
Baccharis halimifolia.
Baccharis (sessile). This shrub is dioecious as is the preceding but it differs principally in that the flowers are sessile, it is only found in very moist areas. The leaves have the same shape as the preceding.
Nyssa aquatica. This tree only grows in submerged areas.

Nyssa montana. This tree is common in moist areas, but rarely forms fruits in dry areas.
Nyssa montana. This is a variety in the northern parts, the wood is much liked by craftsmen as is that of the preceding sp.
Fraxinus (aquatica). It is called swamp ash and it only grows in submerged areas.
Andromeda mariana no. 1.
Andromeda (sempervirens) no. 2. It is Andromeda nitida of Bartram. Very moist areas.
And. racemosa. no. 3.
And. paniculata. no 4. This shrub is not ornamental.
Cyrilla racemiflora. This shrub is the Andromeda plumata of Bartram.
Bignonia radicans.
Bignonia sempervirens.
Bignonia crucigera.
Callicarpa. Sandy and dry soils.
Glycine frutescens. Very moist areas.
Clethra.
Itea.
Amorpha.
Smilax laurifolia no. 1. Very moist areas.
Sm. sarsaparilla? no. 2.
Sm. (laevis) no. 3.
Sm. baccis rubris. no. 4. Submerged area.
Sm. baccis rubris. no. 5. Dry areas. Cissampelos?
Chamaerops (caroliniana) Palmeto.
Gerardia (asparagifolia). Annual ornamental plant.
Lianne-Supple-Jack. Climbing vine from wet areas. It is not named by botanists and I give it a common name.
Vitis? (laciniata).
Pontederia cordata. Aquatic area.
Croton sebiferum. Waxy tree.
Croton argenteum?
Xiris. Aquatic plant.
Unknown tree Daphne? Sandy area, seems to be affected by frost.
Hypericum ericoides. Small annual plant.
Rhexia mariana. Pretty plant.
Halesia tetraptera.
Myrica cerifera arborea.

M. cerifera pumila.
Evonimus sempervirens. This shrub is very pleasant looking in areas that are always moist and shady.
Liquidambar.
Eringium aquaticum. Plant of the rattlesnake.
Mespilus.
Erythrina herbacea.
Viburnum cassinoides. Moist area.
Vib. nudum? Very aquatic areas.
Vib. prunifolium.
Ascyrum. Small shrub.
Phaseolus. Pea. Fruit swordlike.
Milinum. Millet.
Cercis americana. Judas Tree?
Green oak.
Cypressus disticha.[1]
Pavia lutea.[1]
Laurus borbonia.[1]
Vaccininum arboreum.[1]
Juglans hickory.[1]

1. These seeds in the 3rd box should be sown immediately while opening the case in order not to change the moisture that they have absorbed.

2-7. From a copy sent by Marie-Florence Lamaute. Permission from the Bibliothèque Centrale du Muséum National d'Histoire Naturelle, Paris.

Monsieur Le Comte
I have the honor of informing you that I sent via an American ship headed for Le Havre de Grace, and addressed to M. Rolland and Company, 5 ducks of the kind called by the Count De Buffon Beautiful Duck of Louisiana and by the Americans Summer Ducks, Canard & etc. They are also called Wood Ducks because they never are found on the river, where ocean water backs up, but live only on lakes in the middle of the woods. They perch on the trees and make their nests in the trees, probably to avoid being surprised by snakes which are abundant here. I gave to M. Cuvillier all the details concerning them, and I hope that they will arrive at Le Havre in good shape. To get them used to their surroundings I put them on board the ship more than 10 days before the departure.

I'm ready to go to Georgia and even to Florida if I don't experience any difficulties . . . with the Spanish or the Indians. I have been waiting only because I've been busy for several days making the accounts of my expenses. They are finished, and I will send them to you, Monsieur Le Comte by a ship which is leaving in 3 or 4 days for L'Orient.

I sent you 22 boxes of trees by an American ship which left here for Bordeaux. It would have arrived in France by now, but it hit a sand bar when leaving Charleston. The ship was unloaded, and the boxes were exposed for several days to a very ___north wind; I'm afraid the trees may have been damaged. When I arrived at that spot they were unloading the ship. I had the boxes put in a storehouse, but I'm afraid the delay will be a bit harmful.

I was obliged, Monsieur Le Comte, to get from Monsieur Dutartre the sum of [words unclear] remains still some money from the previous amounts, but since there was no other way to obtain money, as one receives here only paper money, I had to submit to this necessity.

I hope to send you, Monsieur Le Comte, by the first boat, the account of how I have used the money I've received, and [I hope to] show you the advantage of the acquisition of land that I've made, both for the savings and for the conservation of the discoveries that are to be made here. I wasn't able to send you the copy of the deed of that purchase because it is housed with the lawyer who did the transaction until the completion of payment, which is supposed to be six months from the purchase.

I am with great profound Respect
Monsieur Le Comte
Your very humble and very obedient servant
Charleston the 8 April 1787 A. Michaux

2-8. Letter from André Michaux to André Thouin. From a copy sent by William Reed of the William Reed Company.

Charleston the 6th of November 1787
Monsieur
I have the honor to announce the shipment of a little package of seeds in a box addressed to Monsieur Le Monnier. I made two other shipments to you on 10 August and on the 25 September of this year. I was very upset to learn of the poor success of the shipment to Monsieur Le Comte d'Angiviller through Bordeaux. This shipment had been made

on 15 February to coincide with the departure of the vessel. It deferred its departure, then it was disabled, the boat was emptied and the boxes underwent another delay of three weeks and were exposed during this time to cold and dry winds.

L'Abbé Nolin complains of the way that I have executed the wrapping of trees, he reprimands me for having enveloped roots of trees with pine needles. One would have to be without sense to work in such a fashion. I used the pine material to separate areas between branches and to prevent fermentation which could happen in the ships if one were to put as much fresh moss in the middle of the box and among the branches as we must put around the roots. In the bottom of the boat where the boxes are usually placed, fermentation and heat often make the trees develop their buds and leaves.

L'Abbé Nolin reproaches me even more about the shipments that the gardener [Saunier] made last winter. Before leaving for Carolina, we selected together a quantity of the most interesting trees and shrubs, I sent the duplicate of this list to l'Abbé Nolin and he approved it. I have placed in this list fewer specimens of each species than the gardener himself promised to send so that he would not complain that I demanded more than he could furnish. In addition to the list I asked him to send nuts of Juglans nigra that I bought myself from a private individual, about 2 miles from the Nursery on the way to New York because this tree which is native to more southern areas is not found in the forests in the state of New York. All he had to do was to obtain the amount needed when the nuts were ripe; this should not have been a pretext for his not collecting other seeds. I made people gather moss two months before my departure. We worked on this task every morning because the snakes are dangerous during the day. A quantity more sufficient than necessary was collected and dried. I had made a number of boxes in order to be ready at the time of shipment. I looked over his herbarium and I named the plants so that he would know the plants that are the most interesting. I told him that I would make interesting shipments and that we would compare my shipments with his and that he should force himself to do better than I. The shrubs that are of most interest to l'Abbé Nolin are Kalmia angustifolia and latifolia, Azalea, Rhododendron, Andromeda [and] Vaccinium. They are all growing on the King's land with the exception of Kalmia latifolia which is found only on the hills two miles away. He did not send any of these shrubs and said the freezes surprised him. He did not send seeds of Tulip Poplar either which was mentioned on the list and much desired by Monsieur Abbé Nolin.

I recommended that he cultivate the area with a spade or cultivate a portion of the lower terrain known to be very favorable to seeds that we proposed. He can hire workmen by the day as there is money deposited for his needs at the consulate in New York. When I arrived this year in August I did not see any seeds, he told me that everything that he sowed died. He used all of the land and probably all his time for cultivating corn, buckwheat, barley, and potatoes. He had a good harvest. I had told him to use for his profit all the land which was not occupied by seeds of trees and I would have seen his harvest with pleasure as I gave him all possible encouragement. He lives in a new house, he finds wood for heating on the King's land without even being obliged to cut for several years. He has pigs and cows which run in neighboring forests like other inhabitants and thus without expense they furnish a part of his food. He takes hay for the maintenance of his animals in winter on the lower portion of the King's land. I had recommended to him seeds that would have given him much honor because it shows the effect of work and specimens of young plants are worth much more than those that are taken in the woods. Even if he had only worked 30 toises [ca. 60 m] with the flat shovel and sowed ca. 2 m with each kind of trees, for example 2 m of tulip poplar, 2 m of Liquidambar, another in Magnolia glauca, another with oaks, etc. This would have furnished more than two or even three hundred for each 2 m if we suppose the success which rarely is lacking if the land is well prepared and if one sows immediately the seeds of the trees. During my stay in New York I did not cease to give him all possible encouragement. This encouragement only served to increase his avarice and rendered him blind to his own interest, as he thinks only of making an immediate profit. At first when I arrived in America seeing him very hard working and sober I had been moved to tell you good things that I noticed, but time has shown his character. If I recommend something to him, he shouts and makes a sermon that I can trust him that he will do the work and I should not even think about it.

I did not complain against him but made only vague remarks to justify my position concerning what he said about the collection of walnuts which he had said kept him from obtaining other seeds. He flatters himself to have your esteem, but as everything that you have written concerning him had as its object the good of the project, I pray please tell me what to do. Knowing the confidence that he has in you and the fear of losing your esteem, I am persuaded that your recommendations and your reprimands will have all the possible success.

These unpleasantnesses but more particularly those stemming from undeserved reproaches from l'Abbé Nolin would be more than enough to cause me to withdraw if they were to continue, because I do not work as I do for the honorarium which is given to me, I do not work either like good souls who do not realize their merit but are still useful. I placed my reward in the honor and merit of having executed my mission, and have put into it the effort of not letting any discoveries [if possible] be made by those who come to visit these lands after me. The approbation of Monsieur Le Comte d'Angiviller is not the least reward of which I was desirous, but also my being named correspondent of the Académie des sciences which is well deserved and which I have already asked for. If thus the hope of these rewards are lost because of the reproaches of l'Abbé Nolin who indicates that I have worked poorly and have lost the esteem of the Comte, will I be compensated by the continued honorarium and by the profits of cultivation done upon the counsel of l'Abbé Nolin in the King's nursery in Carolina? I could not acquire these profits of which I would be ashamed if they were against the interest of my mission.

The observations of my trip to Georgia and the Appalachian, the descriptions of the plants which I collected (they are with my herbarium which I have not had a chance to visit since my return) the probability of obtaining more rich natural historical discoveries in the next year, have given me the right to aspire for this title of corresponding member of the Académie for the publication of my discoveries which are increasing daily. I can cite seven species of Ludwigia, three Rhexia which do not coincide with those of Linnaeus. The descriptions of Laurus aestivalis, L. borbonia, L. indica, etc. must be redone and are not at all recognizable. I would like that France beat the English in publishing these discoveries. There are here two ignorant men sent by nurseries, who pick up everything and send some also to Mr. Banks.

I also had remarks about a collection of birds, for the most part new, which I addressed to Monsieur Daubenton during my trip to New York this year in August. I will increase by many this winter the collection of wood samples. I would like to present my request to Monsieur Le Monnier to have his approbation regarding this title of correspondent to the Académie, but I do not even know if he is receiving my frequent shipments. I think also that he puts me in the rank of those people who are not aware of the worth of their work.

Excuse my sincerity and believe in my affection
Sir and dear friend
Your very humble and very obedient servant
A. Michaux

The forbidding of sending seeds to anyone has kept me from sending those I had promised to Messieurs Cels and L'Heritier.

Wouldn't it be possible to profit from the garden that I have in this climate, which is so favorable for the growth of plants, for increasing seeds of which you are desirous? I would send them to you in abundance and this would only be between the two of us.

2-9. From a copy sent by Marie-Florence Lamaute. Permission from the Bibliothèque Centrale du Muséum National d'Histoire Naturelle, Paris.

Monsieur Le Comte

I have the honor of sending you four cases of trees, two cases, and one box of seeds. Among the trees are two new kinds, namely: a shrub that once was named Aster fruticosus and an Alaterne [Alaternus] new species.

I was preparing to go to Georgia so as to send you this winter that which I found there last summer, but the Indians kept laying waste the land, and I am hastening to take advantage of the permission of the Spanish governor to go to Florida. I am confident in advance about the advantages that will result from this, and I'll let you know about it as soon as possible.

I am with great Profound respect
Monsieur Le Comte
Your very humble and very obedient servant
A. Michaux

From Charleston, the 9 February 1788

There has arrived here, Monsieur Le Comte, an English packet boat according to arrangement which has been made by the English minister for freighters from England to Jamaica. They are supposed to stop at Charleston when returning to Europe, because that stop, which is so advantageous to the Carolinians and to the English, costs little and slows down the boat's return very little.

2-10. From a copy sent by Marie-Florence Lamaute. Permission from the Bibliothèque Centrale du Muséum National d'Histoire Naturelle, Paris.

St. Augustin the 24 April 1788
Monsieur Le Comte

I have the honor to announce my departure from Charleston to Florida. I left the 14th of February and did not arrive until the 28th because of the contrary winds. As soon as I arrived I employed the means to fulfill my objective. I obtained from the Government the permission to travel in all parts of the Province and from the Officers attached to the Government all the information and all the facilities which were at their disposal. I directed first of all my trip in the southern part and wished to go until the Cape of Florida, but as this Area is only inhabited by Indians, I could only advance to Cape Canaveral and a little bit further to 28 degrees 15 minutes Latitude. It would not have been impossible for me to go much further, but seeing that the more I went the more I discovered trees and plants which are Tropical in nature, I was afraid that although there were many novelties, our intentions, Monsieur Le Comte, would not have been fulfilled in using up money and employing much time for trees and plants that were not suited for the French climate. Although I did not collect a large number of seeds, because it was not the season, I cannot send them all at this time—to keep the Package from getting too large that I am leaving with the Spanish Government to be sent to Philadelphia.

I just received notice, Monsieur Le Comte, to transfer funds for the administration of the Establishment in New York and I sent to Monsieur de la Forest a voucher for 2000 francs for Monsieur Dutartre on presentation 30 days hence.

I am with great profound Respect
Monsieur Le Comte
Your very humble and very obedient servant
A. Michaux

2-11. From a copy sent by Marie-Florence Lamaute. Permission from the Bibliothèque Centrale du Muséum National d'Histoire Naturelle, Paris.

Monsieur Le Comte

I have the honor to announce my return from Florida. This trip was a very happy one because of the large number of rare and new Plants which I procured. I received from the Government and from all of the civil and military Officers of this Government a most favorable welcome. I had the honor of writing to you while in St. Augustin [Augustine] and sending you Seeds of this Province by a Ship which was destined for New York.

I cannot at the moment send you, Monsieur Le Comte, the seeds that I brought back here or even give you any details since I have not yet been able to remove them from the Ship on which I arrived here. I am nevertheless obliged to obtain from Monsieur Dutartre the sum of 3000 pounds to the order of Monsieur Jean Jacques Himely within a 30 days notice, a part of this sum is destined to reimburse immediately advances which had been furnished to me while I was in Florida.

I am with great profound Respect
Monsieur Le Comte
Your very humble and very obedient servant
A. Michaux
Charleston 10 June 1788

2-12. From a copy sent by Marie-Florence Lamaute. Permission from the Bibliothèque Centrale du Muséum National d'Histoire Naturelle, Paris.

Monsieur Le Comte

I have the honor to send you a third shipment of the seeds obtained in Florida. The trip which was out of season for seeds produced relatively few in comparison with the different types of interesting and rare trees which I recognized and I will make every effort to send you these in favorable time. I am contemplating another trip of which the success is almost completely assured because it has the objective to collect the discoveries of last year, which I would not be able to execute if I went to New York.

According to the last word that I have received, Monsieur Le Comte, things appear better taken care of, and I hope that Monsieur l'Abbé Nolin will find the shipment from Saulnier more to his liking.

The difficulty of finding funds here is such that Monsieur Petry was not able to place my check of the 1st of January until three months after it had been sent to Philadelphia. Although there are funds left from the 1st check, the circumstance and that of the voyages that will occur soon oblige me to profit from the offer of Monsieur de Leyritz, Officer in the Regiment of Vienna who is returning to France. He proposed to give me funds up to 3000 pounds with a letter of credit for Monsieur Dutartre dated lst of July in Charleston order of Monsieur de Leyritz with a 30 days notice.

Monsieur de Leyritz is very well educated and will be able to give you, Monsieur Le Comte, the particulars regarding this country here.

I am with great profound Respect

Monsieur Le Comte

Charleston 1 July 1788 Your very humble and very obedient servant

A. Michaux

2-13. From a copy sent by Marie-Florence Lamaute. Permission from the Bibliothèque Centrale du Muséum National d'Histoire Naturelle, Paris.

Monsieur Le Comte

I have the honor of sending you a little Box of the first seeds gathered this year. I added to it those that remained from the trip to Florida, of which I sent you two preceding shipments.

I hope Monsieur Le Comte to send you soon many new trees and seeds from these same trees. The war with the Indians of Georgia is over, and we'll leave before the end of the Month for the Harvest of these.

I am with great profound Respect

Monsieur Le Comte

Your very humble and very obedient servant, Charleston the 2 August 1788

A. Michaux

List of Seeds Sent 2 August 1788

Fothergilla gardeni
Chionanthus virginicus
Magnolia glauca
Viburnum dentatum
Aletris farinosa
Podophyllum peltatum

Helonias asphodeloides
Malva carolinina
Laurus sassafras
Conyza virgata

———

Guilandina bonduncella
Sophora occidentalis
Chiococca racemosa
(frutex ericaefolia)
Ceanothus (floridanus)
Conocarpus racemosa
Coffea? new species
Hippomane
Amorpha
Unknown shrub
Amyris elemifera

2-14. From a copy sent by Marie-Florence Lamaute. Permission from the Bibliothèque Centrale du Muséum National d'Histoire Naturelle, Paris.

Monsieur Le Comte
I have the honor of sending you a box and two cases of seeds. I only received eight days ago the news of the discontinuation of the Ships [from New York to L'Orient]. Because of this I only made up one little box of well chosen seeds and I will send larger shipments by boats which leave here directly for France. I am assured that at the end of December there will be merchant ships for Le Havre.
I am with very deep Respect
Monsieur Le Comte
Your very humble and very obedient servant
Charleston the 1st October 1788 A. Michaux

List of Seeds Sent the 1st of October 1788

No. 1 Box recommended for Monsieur de la Forest

Fothergilla gardeni
Styrax angustifolia
Magnolia glauca
Magnolia tripetala
Podophyllum peltatum

Helonias asphodeloides
Aletris farinosa
Hibiscus . . .
Gerardia lutea
Conyza virgata
Alaternum . . .
Pinus foliis longissimis
Clematis viorna
Tilia (caroliniana)
Zanthoxilum clava herculi
Halesia tetraptera
Chionanthus virginicus

No. 2 One case of Fagus pumila } Chinquapin
No. 3 One case of the same Fagus pumil

These two cases which cannot stand a long delay will stay in New York; if there is no ship ready to leave for Le Havre then the seeds will be sown.

2-15. From a copy sent by Marie-Florence Lamaute. Permission from the Bibliothèque Centrale du Muséum National d'Histoire Naturelle, Paris.

Monsieur Le Comte
I have the honor of sending you eight cases of seeds and two cases of trees. I sent it on an American ship destined for Le Havre de Grace and recommended by the tradesman, Mr. Limousin. The man in charge of loading the boat at first had refused, saying that the boat was already fully loaded with its own merchandise and that he would not accept another load, but having with Mr. Petry presented the fact that this shipment was in the service of the King, we obtained permission to send six cases. Then seeing the good nature of the loader, I presumed that I could add a case of trees for Monsieur and two little boxes of seeds. These two boxes hold in small quantity the same species found in the eight large cases. One is destined for Monsieur and the other for the Jardin du Roi, if you agree with these destinations, Monsieur Le Comte.

The cessation of packet boats having made the shipping less frequent, I do my utmost to make them more interesting and one of the two cases of trees contains entirely new species that have not been sent before.

I am with very deep respect

Monsieur Le Comte

Your very humble and very obedient servant

Charleston the 5 January 1789 A. Michaux

(Authors' comment: Michaux gives two nearly identical lists for January 5, 1789. We have combined the two lists into one presented below.)

List of Trees and Seeds sent 5 January 1789

Seeds

First Case
Cupressus disticha

2nd Case
Cupressus disticha

3rd Case
Cupressus disticha

4th Case
Liriodendron tulipifera

5th Case
Pinus foliis longissimis

6th Case
Collection of several kinds:
Andromeda mariana
And. racemosa
And. paniculata
And. nitida (And. lucida)
Baccharis halimifolia
Bignonia catalpa
Bign. crucigera
Bign. sempervirens
Clethra alnifolia
Conyza virgata
Halesia tetraptera
Styrax angustifolia

Chionanthus virginicus
Fraxinus americana
Nyssa aquatica
Nyssa dentata
Ziziphus scandens
Alaternus caroliniana with large leaves
Magnolia glauca
Magn. tripetala
Diospiros virginiana
Itea virginica
Kalmia latifolia
Hypericum kalmianum
Cyrilla racemiflora
Hamamelis virginiana
Fothergilla gardeni
Stewartia malaccodendron
Gossypium
Myrica cerifera
Annona grandiflora
Convolvulus dissectus

7th Case
Acorns of three species of oaks:
Willow oak
Red oak
White oak

8th Case
Acorns of green oak or live oak
Chinquapin

Trees

9th Case
60 Calycanthus floridus, wet and sandy soils
30 Ilex angustifolia, very wet areas
3 Annona grandiflora
2 Andromeda arborea, tree of 4 or 5 feet in circumference; sandy and cool soils
8 Carpinus with fruit resembling hops
50 Betula papyrifera
8 Pavia, flowers in spike

25 Unknown shrub with leaves of Erica scoparia and berries with two seeds
2 Ceanothus new species
12 Rhododendron new species
20 Epigea repens
10 Illicium
50 Spigelia marylandica

10th Case
30 Styrax latifolia (also listed as Styrax angustifolia)
14 Alaternus with large leaves (Alaternus caroliniana)
30 Nyssa dentata, in very wet places
15 Magnolia grandiflora
9 Magn. glauca
52 Sideroxilum tonex [Sideroxylon tenax]
6 Ilex, in wet places
225 Chinquapins
20 Lirodendron tulipifera
125 Cypressus disticha
12 Gleditsia monosperma or aquatica
30 Rhododendron new species, very young Plants

No. 11: Box of seeds for the Jardin du Roi

2-16. From a copy sent by Marie-Florence Lamaute. This letter is probably to the King's brother, to whom Michaux referred as "Monsieur." Permission from the Bibliothèque Centrale du Muséum National d'Histoire Naturelle, Paris.

Charleston the 5th January 1789
received 28 February 1789
Monsieur

I am sending you under the address of Monsieur Le Comte D'Angiviller a box of seeds from Caroline in which there is also a little package of seeds from Florida that [it] had not been possible to send earlier upon my return from this trip last year.

I have the honor of being
Sir
Your humble and very obedient servant
A. Michaux

List of seeds in box No. 11 addressed to Monsieur Le Comte D'Angiviller

from Caroline

Andromeda mariana
A. paniculata
A. racemosa
A. lucida
Baccharis dioecious
Bignonia sempervirens
Conyza virgata
Halesia tetraptera
Styrax angustifolia
Fraxinus americana
Nyssa aquatica
N. dentata
Ziziphus scandens
Alaternus
Magnolia glauca
M. tripetala
Diospiros
Clethra
Itea
Kalmia latifolia
Cyrilla
Hamamelis
Fothergilla
Stewartia
Annona grandiflora
Convolvulus dissectus
Gaura biennis
Cypressus disticha
Pinus fol. longissimis
Chionanthus
Myrica cerifera
Chironia new species
Kuhnia ?
Viburnum dentatum
Liriodendron tulipifera
Amaranthus 20 feet high, dioecious

Seeds from Florida
Sophora occidentalis
Unknown tree
Unknown shrub with opposite leaves
Unknown climbing shrub
Guilandina bonducella
Dolichos shrub
Hippomane? herbaceous
Zamia pumila
Ceanothus new species
Cacalia? shrub
Shrub bearing berries and having leaves of Erica
Eriocaulon
Conocarpus racemosa
Tillandsia lingulata
Malva caroliniana
Bombax gossypinum
Chiococca racemosa
Coffea?
Erythrina arborea
In a case of trees addressed to Monsieur apart from the six Illicium, there are two in addition on top of the other trees that Monsieur Thouin can give to Monsieur Lemonier.

2-17. From a copy sent by Marie-Florence Lamaute. Permission from the Bibliothèque Centrale du Muséum National d'Histoire Naturelle, Paris.

Monsieur Le Comte
I have the honor of sending you seeds gathered in my trip to the Bahamas. I recognized there all the trees cited by Catesby, with the exception of two or three species only; a very large number of these have been described by Plumier, Jacquin and Sloane in the Antilles; and several new species. I brought back more than 1500 trees which I planted, waiting for the right season to send them to you. I used only two months for this trip so as not to miss the season to go to the mountains of North Carolina. These mountains are the highest of all those which make up the Allegheny and Appalachian, and their products can succeed better in France than those of the other southern parts of the U.S. The temperature there is so moderate that I found in the mountains of South Carolina trees of Canada and of New York state.

I just received two wild turkeys, though I have been asking for them all along; now I have more hope of receiving more before next autumn.

I am with great profound Respect

Monsieur Le Comte

Your very humble and very obedient servant

Charleston the 28 April 1789 A. Michaux

List of Seeds [from the Bahamas] sent 28 April 1789

Amyris elemifera
Catesbaea spinosa
Lytisus cajan
Duranta plumieri
Vinca lutea
Pisonia aculeata
Iresine celosioides
Swietenia mahagoni
Rauwolfia glabra
Mimosa circinalis
Chrysophyllum . . .
Carica papaya
Malva abutiloides
Hibiscus manihot
Turnera sidoides
Chrysocoma n. sp.
Helicteres isora
Passiflora rubra
Passifl. laurifolia
Passifl. cuprea
Passifl. suberosa
Passifl. incarnata

2-18. André Michaux's request to Governor Zéspedes to make observations in the East Florida Province, March 8, 1788, St. Augustine (East Florida Papers, 51). Permission from the Director General del Libro, Archivo y Bibliotecas, Archivo General de Indias, Seville.

Mr. Governor and Captain General
No. 158
Mr. André Michaux, subject of His Most Catholic Majesty and Professor of the Botanical Sciences, respectfully informs you of his desire to conduct botanical observations in this Province which undoubtedly will lead to the advancement of botanical knowledge.

I respectfully request your Excellency that you grant me the authority to travel in the Province accompanied by two assistants and a Negro [word unclear]. They will accompany me to verify the proposed observations. Your kindness in granting me this favor will eternally be remembered. I am hopeful to make a discovery which will be worthy of bearing your name as a symbol of my eternal respect for you. St. Augustin [Augustine] of Florida, 8 March 1788
André Michaux

Literature Cited

Adams, W., D. Schafer, R. Steinbach, and P. Weaver. *Kings Road—Florida's First Highway*. St. Augustine, Fla.: Historic Property Associates, 1997.

Ambrose, S. E. *Undaunted Courage*. New York: Touchstone, Simon & Schuster, 1996.

Beale, G. R. "Bosc and the Exequatur." *Prologue, the Journal of the National Archives* 10, no. 3 (1978): 133–151.

Berkeley, E., and D. S. Berkeley. *Dr. John Mitchell: The Man Who Made the Map of North America*. Chapel Hill: University of North Carolina Press, 1974.

———. *The Life and Travels of John Bartram: From Lake Ontario to the River St. John*. Tallahassee: University Presses of Florida, 1982.

Brewer, D. M. "Archeological Overview and Assessment Canaveral National Seashore." Unpublished report. Southeast Archaeology Center, National Park Service, Tallahassee, Fla., 1988.

Brunet, O. *Voyage d'André Michaux en Canada depuis le Lac Champlain jusqu'à la Baie d'Hudson*. Québec: Bureau de l'Abeille, 1861.

———. "Notice sur les Plantes de Michaux et sur Son Voyage au Canada et à la Baie d'Hudson d'après Son Journal Manuscrit et autres Documents Inédits." *American Journal of Science* 37 (1864): 276–287.

Catanzariti, J. (ed.). *The Papers of Thomas Jefferson*, vol. 25: 1 January to 10 May 1793. Princeton, N.J.: Princeton University Press, 1992.

Catesby, M. *The Natural History of Carolina, Florida and the Bahama Islands*, vol. 2. London: Printed for B. White, 1771.

Cheston, E. R. *John Bartram, 1699–1777. His Garden and His House*, 2nd ed. Philadelphia: John Bartram Assoc., 1953.

Chinard, G. "André and François André Michaux and Their Predecessors. An Essay on Early Botanical Exchanges between America and France." *Proceedings of the American Philosophical Society* 101, no. 4 (1957): 344–361.

Coats, A. M. *The Quest for Plants. A Study of the Horticultural Explorers*. London: Studio Vista, 1969.

Coker, W. C. "The Garden of André Michaux." *Journal of the Elisha Mitchell Scientific Society* 27, no. 2 (1911): 65–72.

Coker, W. S., and S. R. Parker. "The Second Spanish Period in the Two Floridas," pp. 150–166. In: M. Gannon (ed.), *The New History of Florida*. Gainesville: University Press of Florida, 1996.

Coker, W. S., and T. D. Watson. *Indian Traders of the Southeast Spanish Borderlands: Panton, Leslie & Company and John Forbes & Company*, 1783–1847. Pensacola: University of West Florida Press, 1986.

Core, E. L. "The American Species of Scleria." *Brittonia* 2, no. 1 (1936): 1–105.

Corse, C. D. *Dr. Andrew Turnbull and the New Smyrna Colony of Florida*, rev. ed. St. Petersburg, Fla.: Great Outdoors, 1967.

Coulter, J. M. "Some North American Botanists. III. André Michaux." *Botanical Gazette* 8, no. 3 (1883): 181–183.

Daudin, F. M. *Histoire Naturelle, Générale et Particulière, des Reptiles*, vol. 2. Paris: L'imprimerie de F. Dufart, 1802.

Deleuze, J. P. F. "Notice Historique sur André Michaux." *Annales du Muséum National d'Histoire Naturelle* 3 (1804): 191–227.

———. "Notice Historique sur André Michaux." *L'Hemisphère, Journal Littéraire et Politique* 1, no. 14 (1810): 209–215.

Derr, M. *Some Kind of Paradise: A Chronicle of Man and the Land in Florida*. Gainesville: University Press of Florida, 1998.

Dewhurst, W. W. *The History of Saint Augustine, Florida*. 1885. Reprint, Rutland, Vt.: Academy Books, 1968.

Dorr, L. J. *Plant Collectors in Madagascar and the Comoro Islands*. London: Royal Botanical Gardens, 1997.

Dovell, J. E. *Florida: Historic, Dramatic, Contemporary*, 4 vols. New York: Lewis, 1952.

Duprat. G. "Essai sur les Sources Manuscrites Conservées au Muséum National d'Histoire Naturelle," pp. 231–252. *In*: J. F. Leroy (ed.), *Les Botanistes Français en Amérique du Nord avant 1850*. Paris: Centre National de la Recherche Scientifique, 1957.

Dutilly, A., and E. Lepage. "Retracing the Route of Michaux's Hudson Bay Journey of 1792." *Revue de l'Université d'Ottawa* 15, no. 1 (1945): 89–102.

East Florida Papers 51. Section 44. 1788 March 8, EF r1, 77 bnd 179J14. Document 1788–3. Letter of André Michaux to Governor Zéspedes.

East Florida Papers 58. Section 69–71. 1797 January 26, EF r1, 137 Document 1797/14. Jacob Wiggins Testamentary proceedings, 295 pp. Hist. Records Surv. Div., Spanish Land Grants in Florida, 1941.

Eifert, V. S. *Tall Trees and Far Horizons: Adventures and Discoveries of Early Botanists in America*. New York: Dodd, Mead, 1965.

Ewan, J. "French Naturalists in the Mississippi Valley," pp. 159–174. *In*: J. F. McDermott (ed.), *The French in the Mississippi Valley*. Urbana: University of Illinois Press, 1965.

——— (ed.). *A Short History of Botany in the United States*. New York: Hafner, 1969.

Fabel, F. A. "British Rule in the Floridas," pp. 134–149. *In*: M. Gannon (ed.), *The New History of Florida*. Gainesville, University Press of Florida, 1996.

Forbes, J. G. *Sketches, Historical and Topographical, of The Floridas; More Particularly of East Florida*. New York: C. S. Van Winkle, 1821.

Gannon, M. V. *The Cross in the Sand: The Early Catholic Church in Florida, 1513–1870*. Gainesville: University of Florida Press, 1965.

———. *Florida: A Short History*. Gainesville: University Press of Florida, 1993.

Gillmore, Q. A. "Plan and Estimate for Opening a Passage Between the North End of Indian River and Mosquito Lagoon, Florida." 48th Congress 1st Session, Senate, Ex. Doc. No. 65, pp. 8–13. United States Engineer Office, New York, 29 September 1883.

Gray, A. "Remarks Concerning the Flora of North America." *American Journal of Science* 24, no. 143 (1882): 321–331.

Griffin, J. W., and J. J. Miller. "Cultural Resource Reconnaissance of Merritt Island National Wildlife Refuge." Unpublished report. Cultural Resource Management Inc., Tallahassee, Fla., 1978.

Griffin, P. C. *Mullet on the Beach: The Minorcans of Florida, 1768–1788*. Jacksonville: University Presses of Florida, 1991.

Guillaumin, A., and V. Chaudun. "L'Introduction en France des Plantes Horticoles Originaires de l'Amérique du Nord avant 1850," pp. 123–135. *In*: J. F. Leroy (ed.), *Les Botanistes Français en Amérique du Nord avant 1850*. Paris: Centre National de la Recherche Scientifique, 1957.

Hann, J. H. "The Missions of Spanish Florida," pp. 178–99. *In*: M. Gannon (ed.), *The New History of Florida*. Gainesville: University Press of Florida, 1996.

Hanna, A. J., and K. A. Hanna. *Florida's Golden Sands*. Indianapolis: Bobb-Merrill, 1950.

Harper, F. (ed.). "Diary of a Journal Through the Carolinas, Georgia, and Florida from July 1, 1765, to April 10, 1766," by John Bartram. Annotated. *Transactions of the American Philosophical Society*, n.s., 33, pt. 1 (1942): 1–120.

——— (ed.). "Travels in Georgia and Florida, 1773–74: A Report to Dr. John Fothergill," by William Bartram. Annotated. *Transactions of the American Philosophical Society*, n.s., 33, pt. 2 (1943): 121–242.

——— (ed.). *The Travels of William Bartram*, Naturalist's Edition. New Haven, Conn.: Yale University Press, 1958.

Hetrick, L. "The Origins, Goals, and Outcomes of John Bartram's Journey on the St. Johns River, 1765–66," pp. 1–13. Unpublished MS, 2000, in possession of the author.

Hicks, J. D. *The Federal Union*, 3rd ed. Cambridge, Mass.: Houghton Mifflin, 1957.

Higgs, C. D. "Appendix A: Derrotero of Alvara Mexia, 1605," pp. 269–273. *In*: I. Rouse, *A Survey of Indian River Archaeology, Florida*. Yale University Publications in Anthropology no. 44. New Haven, Conn.: Yale University Press, 1951.

Historical Records Survey, Division of Community Service Programs, Work Projects Administration. *Spanish Land Grants in Florida*, vol. 5. Confirmed Claims: S-Z. Pages 3–4; 210–212. Tallahassee, Fla.: State Library Board, May 1941.

Hitchcock, A. S. "Types of American Grasses: A Story of the American Species of Grasses Described by Linnaeus, Gronovius, Sloane, Swartz, and Michaux." *Contributions from the United States National Herbarium* 12 (1908): 113–158.

Hochreutiner, B. P. G. "Validity of the Name Lespedeza." *Rhodora* 36 (1934): 390–392.

Hooker, W. J. "On the Botany of America." *American Journal of Science* 9 (1825): 263–284.

Hunt, K. W. "The Charleston Woody Flora." *American Midland Naturalist* 37, no. 3 (1947): 670–756.

IDC. Microfiche set of André Michaux's herbarium. IDC 6211, 145 microfiches. InterDocumentation Co., Zug, Switzerland, 1968.

Johnson, S. "The Spanish St. Augustine Community, 1784–1795: A Reevaluation." *Florida Historical Quarterly* 68, no. 1 (1989): 27–54.

Joyce, D. D. "Preliminary Report on the Archaeological Investigation of the André Michaux Site. 38CH2022." Unpublished report. College of Charleston, 1988: 1–25, Appendix I, II.

Kastner, J. A. *Species of Eternity*. New York: Alfred A. Knopf, 1977.

Kral, R. "A Revision of *Asimina* and *Deeringothamus* (Annonaceae)." *Brittonia* 12, no. 4 (1960): 233–278.

Lacroix, A. "André Michaux (1746–1803)," pp. 273–322. *In*: A. Lacroix (ed.), *Figures de Savants*, vol. 4. Paris: Gauthier-Villars, Imprimeur-editeur, 1938.

Lamaute, M.-F. "André Michaux et Son Exploration en Amerique du Nord, de 1785 a 1796, D'apres Les Sources Manuscrites." M. A. thesis, Université de Montreal, 1981.

Letouzey, Y. *Le Jardin des Plantes à la Croisée des Chemins avec André Thouin 1747–1824*. Paris: Muséum National d'Histoire Naturelle, 1989.

Lisitzky, G. *Thomas Jefferson*. New York: Viking, 1933.

Lockey, J. B. *East Florida 1783–1785: A File of Documents Assembled, and Many of Them Translated*. Berkeley and Los Angeles: University of California Press, 1949.

Michaux, A. "Mémoire Abrégé Concernant Mes Voyages dans l'Amérique Septentrionale." 22 May 1793. Library of Congress.

———. *Histoire des Chênes de l'Amérique*. Paris: L'Imprimerie de Crapelet, 1801.

———. *Flora Boreali-Americana*, 2 vols. 1803. Reprint, with an introduction by Joseph Ewan, New York: Hafner, 1974.

Michaux, F. A. *Voyage à L'ouest des Monts Alleghanys*. Paris: L'Imprimerie de Crapelet, 1804.

———. *The North American Sylva; or a Description of the Forest Trees of the United States, Canada, and Nova Scotia*, 3 vols. Philadelphia, 1817–18.

Milanich, J. T. *Florida Indians and the Invasion from Europe*. Gainesville: University Press of Florida, 1995.

———. "Original Inhabitants," pp. 1–15. *In*: M. Gannon (ed.), *The New History of Florida*. Gainesville: University Press of Florida, 1996.

Morton, C. V. "The Fern Herbarium of André Michaux." *American Fern Journal* 57 (1967): 166–182.

Mowat, C. L. *East Florida as a British Province, 1763–1784*. 1943. Reprint, with editorial preface by R. W. Patrick, Gainesville: University of Florida Press, 1964.

Norman, E. M. "An Analysis of the Vegetation at Turtle Mound." *Florida Scientist* 39, no. 1 (1976): 19–31.

Nuttall, T. "A Catalogue of the Collection of Plants Made in East-Florida, during the Months of October and November, 1821, by A. Ware." *American Journal of Science and Arts* 5 (1822): 286–304.

Panagopoulos, E. P. *New Smyrna: An Eighteenth-Century Greek Odyssey*. Gainesville: University of Florida Press, 1966.

Parker, S. R. "Above Ground Historical Structures of Canaveral National Seashore." Unpublished manuscript, 1998. In files of Canaveral National Seashore.

———. "Men Without God or King: Rural Settlers of East Florida, 1784–1790." *Florida Historical Quarterly* 69, no. 2 (1990): 135–155.

Porter, C. M. "Philadelphia Story. Florida Gives William Bartram a Second Chance." *Florida Historical Quarterly* 71, no. 3 (1993): 310–323.

Radford, A. E., H. E. Ahles, and C. R. Bell. *Manual of the Vascular Flora of the Carolinas*. Chapel Hill: University of North Carolina Press, 1968.

Rasico, P. D. "The Minorcan Population of St. Augustine in the Spanish Census of 1786." *Florida Historical Quarterly* 66 (1987): 160–184.

Rehder, A. "Michaux's Earliest Note on American Plants." *Journal of the Arnold Arboretum* 4, no. 1 (1923): 1–8.

Rembert, David H., Jr. "The Carolina Plants of André Michaux." *Castanea* 44, no. 2 (1979): 65–80.

Reveal, J. L. *Gentle Conquest: The Botanical Discovery of North America with Illustrations from the Library of Congress*. Washington, D.C.: Starwood, 1992.

Reveal, J. L., and J. S. Pringle. "Taxonomic Botany and Floristics," pp. 157–192. *In*: Flora of North America Editorial Committee (eds.), *Flora of North America North of Mexico*, vol. 1. New York: Oxford University Press, 1993.

Rey, L. "Deux Botanistes Français aux Etats-Unis: Les Missions de Michaux Père et Fils (1785–1808)." Paris: Bibliothèque du Muséum National d'Histoire Naturelle, 1954.

Ricker, P. L. "The Origin of the Name Lespedeza." *Rhodora* 36 (1934): 130–132.

Robbins, W. J., and M. C. Howson. "André Michaux's New Jersey Garden and Pierre Paul Saunier, Journeyman Gardener." *Proceedings of the American Philosophical Society* 102, no. 4 (1958): 351–370.

Romans, B. *A Concise Natural History of East and West Florida*. 1775. Reprint, with an introduction by R. W. Patrick, Gainesville: University of Florida Press, 1962.

Rohwer, J. G. "Lauraceae: *Nectandra*." *Flora Neotropica* 60: (1993): 1–332.

Ross, C. A., and E. H. Ernst. *Alligator mississippiensis*, pp. 600.1–600.14. *In* Herpetological Catalogue Committee (eds.), *Catalogue of American Amphibians and Reptiles*. Bethesda, Md.: American Society of Ichthyologists and Herpetologists, 1994.

Rousseau, J. "Le Voyage d'André Michaux au Lac Mistassini en 1792." *Mémoire du Jardin Botanique de Montréal* 3 (1948): 1–34.

Rusby, H. H. "Michaux's New Jersey Garden." *Bulletin of the Torrey Botanical Club* 11, no. 8 (1884): 88–90.

Sargent, C. S. "Portions of the Journal of André Michaux, Botanist, Written During His Travels in the United States and Canada, 1785 to 1796. With an Introduction and Explanatory Notes." *Proceedings of the American Philosophical Society* 36, no. 129 (1889): 1–145.

Sastre, C. "Picolata on the St. Johns: A Preliminary Study." *El Escribano* 32 (1995): 25–64.

Sauer, J. "Revision of *Canavalia*." *Brittonia* 16 (1964): 106–181.

Savage, H., Jr., and E. J. Savage. *André and François André Michaux*. Charlottesville: University of Virginia Press, 1986.

Schafer, D. L. "Early Plantation Development in British East Florida." *El Escribano* 19 (1982): 37–53.

———. "'the forlorn state of poor Billy Bartram': Locating the St. Johns River Plantation of William Bartram." *El Escribano* 32 (1995): 1–11.

Schoepf, J. D. *Travels in the Confederation [1783–1784]*. Edited and translated by A. J. Morrison. Philadelphia: William J. Campbell, 1911.

Seaborn, M. M. (ed.). *André Michaux's Journeys in Oconee County, South Carolina, in 1787 and 1788*. Walhalla, S.C.: Oconee County Library, 1976.

Shannon, M. "An Eye for the New: In Praise of André Michaux." *Atlanta Weekly*, 15 May 1983: 26–28.

Siebert, W. H. *Loyalists in East Florida 1774–1785*, vols. 1 and 2. DeLand: Florida State Historical Society, 1929.

———. "Slavery and White Servitude in East Florida, 1726–1776." *Florida Historical Quarterly* 10, no. l (1931): 3–23.

Slaughter, T. P. *The Natures of John and William Bartram*. New York: Alfred A. Knopf, 1996a.

———. *William Bartram: Travels and Other Writings*. New York: Library of America, 1996b.

Spongberg, S. A. *A Reunion of Trees*. Cambridge: Harvard University Press, 1990.

Stafleu, F. A., and R. S. Cowan. *Taxonomic Literature*, vol. 3, 2nd ed. Utrecht: Bohn, Scheltema and Holkema, 1981.

Stevens, P. F. *The Development of Biological Systematics: Antoine-Laurent de Jussieu, Nature, and the Natural System*. New York: Columbia University Press, 1994.

Strickland, A. "Ponce de León Inlet." *Florida Historical Quarterly* 43, no. 3 (1965): 244–261.

Sturtevant, W. "The Southeast," pp. 12–35. In: R. Collins (ed.), *The Native Americans: The Indigenous People of North America*. London: Salamander Books, 1991.

Sunderman, J. F. (ed.). *Journey into Wilderness: An Army Surgeon's Account of Life in Camp and Field during the Creek and Seminole Wars, 1836–1838*, by J. R. Motte. Gainesville: University of Florida Press, 1953.

Tanner, H. H. *Zéspedes in East Florida 1784–1790*. Coral Gables, Fla.: University of Miami Press, 1963.

Tebeau, C. *A History of Florida*, rev. ed. Coral Gables, Fla.: University of Miami Press, 1980.

Thwaites, R. G. (ed.). *Early Western Travels, 1748–1846*, vol. 3. 1904. Reprint of *Journal of Travels into Kentucky; July 15, 1793–April 11, 1796. Andre Michaux*. New York: AMS Press, 1966.

Torrey, J., and A. Gray. *A Flora of North America*, vols. 1 & 2. 1838–1843. Reprint, New York: Hafner, 1969.

Troxler, C. W. "Refuge, Resistance, and Reward: The Southern Loyalists' Claim on East Florida." *Journal of Southern History* 55, no. 4 (1989): 563–596.

True, R. H. "François André Michaux, the Botanist and Explorer." *Proceedings of the American Philosophical Society* 78, no. 2 (1937): 313–327.

Tyler-Whittle, M. S. *The Plant Hunters*. New York: Lyons and Burford, 1997.

Uttal, L. J. "The Type Localities of the *Flora Boreali-Americana* of André Michaux." *Rhodora* 86, no. 845 (1984): 1–66.

Vanstory, B. *Georgia's Land of the Golden Isles*, new ed. Athens: University of Georgia Press, 1981.

Ventenat, É. P. *Descriptions des Plantes Nouvelles et Peu Connues, Cultivées dans le Jardin de J. M. Cels*. Paris: L'Imprimerie de Crapelet, 1800–1803.

Watson, W. H. "Notes and Comment." *Florida Historical Quarterly* 6, no. 2 (1927): 120–29.

Wunderlin, R. P. *Guide to the Vascular Plants of Florida*. Gainesville: University Press of Florida, 1998.

Wyman, J. "Fresh-Water Shell Mounds on the St. John's River, Florida." *Peabody Academy of Science Memoirs* 4 (1875): 3–94.

Zéspedes, V. M. Corresponde a Represent. on Reserbada No. 17. "Descripcion de la Florida Oriental su Clima, Terreno, Productos, Rios, Barras, Bahias, Riertos, Numero, y Calidades de Gente que la habitan &c." Report dated 15 April 1787 to José de Gálvez. Archivo General de Indias, Seville, Spain.

Index

Walter Kingsley Taylor is professor emeritus of biology at the University of Central Florida and the author of *The Guide to Florida Wildflowers* (1992) and *Florida Wildflowers in Their Natural Communities* (UPF, 1998).

Eliane M. Norman is professor emerita of biology at Stetson University, Deland, Florida, and the author of a monograph on New World Buddlejaceae (2000).

Printed in the USA
CPSIA information can be obtained
at www.ICGtesting.com
CBHW020754190924
14316CB00013B/2

9 780813 080451

"*André Michaux in Florida* is outstanding for its scholarship and documentation; the authors have clearly written with meticulous care to detail. . . . A critical documentation of Michaux's botanical work in Florida that adds significantly to the historical record."—***Plant Science Bulletin***

"This book makes a bold start at restoring the significance of [Michaux's] botanical record to the history of science."—***Isis***

"Taylor and Norman have provided a volume useful to specialists on Florida's historical biological diversity. The volume will also be of use to scholars of the late colonial period in Florida, especially those interested in environmental history."
—**H-Net**

"[The authors] provide copious historical, geographic, and biographical background information concerning Michaux, other naturalists with whom he corresponded, and the political history of the area of Florida he explored."—***Journal of Southern History***

"An excellent primary resource for understanding the second Spanish period of Florida."—***Colonial Latin American Historical Review***

THE NAME MICHAUX often appears in the plant names of Florida, from the endangered yellow violets that grow wild in the panhandle to the Florida rosemary of the scrub. André Michaux (1746–1803) was an extraordinary and dynamic individual who explored North America during the eighteenth century, the first trained botanist to explore extensively the wilderness east of the Mississippi River, including Spanish East Florida. This first book-length account of Michaux's Florida exploration combines his original journal with writings about him by later authors, historical background, and the author's own narrative to create a multifaceted, comprehensive treatise on Michaux's travels and discoveries in Florida.

Walter Kingsley Taylor is professor emeritus of biology at the University of Central Florida. **Eliane M. Norman** is professor emerita of biology at Stetson University.

Front: Loblolly bay (*Gordonia lasianthus*) from *Traité des Arbres et Arbustes que l'on cultive en France en pleine terre* (1801–1819) by Pierre-Joseph Redouté. Original from the New York Public Library. Digitally enhanced by rawpixel; licensed under CC BY 2.0.

University Press of Florida
http://upress.ufl.edu

ISBN 978-0-8130-8045-1 $28.00